FORTRAN AT THE KEYBOARD

AN INTRODUCTION TO FORTRAN FOR ENGINEERS

b y

Richard Devon & Wu-chun Feng

College of Engineering

The Pennsylvania State University

KENDALL/HUNT PUBLISHING COMPANY
2460 Kerper Boulevard P.O. Box 539 Dubuque, Iowa 52004-0539

Credits

Appendix A: *WATCOM FORTRAN User's Guide for the IBM PC with DOS,* Chapter 3, pp. 30–48, © 1985 by WATCOM Publications Limited, ISBN 0–919884–14–8. Reprinted with permission of WATCOM Publications Limited.

Appendix B: *WATCOM PC GRAPHICS,* © September 1984 by WATCOM Systems Inc. Reprinted with permission of WATCOM Publications Limited.

Appendix C: *WATCOM EDITOR,* © 1985, 1987 by WATCOM Systems Inc. Reprinted with permission of WATCOM Publications Limited.

Appendix D: *WATCOM FORTRAN,* © September 1984 by WATCOM Systems Inc. Reprinted with permission of WATCOM Publications Limited.

This edition has been printed directly from camera ready copy.

ISBN 0–8403–5461–4

Printed in the United States of America
10 9 8 7 6 5 4 3 2 1

TABLE OF CONTENTS

PREFACE

This text is designed to introduce engineering and science students to FORTRAN in a hands-on, self-paced, friendly microcomputer environment. It has been written with reference to the WATCOM interpreter FORTRAN running on MS-DOS microcomputers. It would be possible to use the text with a few adaptations as an introduction to FORTRAN in *any* environment, with the exception of the last three chapters. These chapters are all heavily dependent on the library graphics subroutines that are intrinsic to the WATCOM interpreter. But, even here, adaptation to FORTRAN compilers linked to other software for the graphics should not be too difficult.

The WATCOM interpreter is particularly attractive for teaching an introduction to FORTRAN. This is because of the easy availability of the graphics subroutines, a good editor, and the lack of a rationale for compiling programs that are very short and can be expected to be particularly error-prone.

Since this is an introductory text, it is not claimed to be comprehensive. A glance at the Table of Contents will reveal that all of the most basic topics are included. However, a number of others are not, such as writing to and reading from files, the use of COMMON statements and blocks, and arrays with more than two dimensions. There has been an explicit attempt not to overwhelm the student with theoretical distinctions and a large volume of material. On the contrary, the student is expected to go to work immediately with each new set of commands and techniques as she/he learns them.

Although the curriculum is clearly learner-oriented, it would be incorrect to assume that the material is simplified and unchallenging. Conceptually, the level is appropriate for the freshman or sophomore year of college, although it could also be used for able-minded high school students.

To a large extent, we have removed the instructor from the curriculum. *Learning* rather than *teaching* is stressed. Teaching a new language tends to over-complicate the material, uses up valuable time in an inefficient way, and bores the instructor. By letting students learn at their own pace and with continuous hands-on practical experience, a better and more efficient learning environment is created. The instructor will only have to give a few short explanatory lectures as the need arises and help the occasional student who has a problem. The student assignments may be graded directly on-screen, another nice feature of this approach. And this book, incidentally, will actually fit conveniently between the keyboard and the monitor.

The curriculum has been used for several years with considerable success at The Pennsylvania State University. Every year approximately 1,000 first year engineering students work through this curriculum. The efficiency of the curriculum may be inferred from the fact that the students complete it in 16-20 hours for about two-thirds of a semester credit. A one semester credit course would be a very comfortable pace.

ACKNOWLEDGEMENT

The authors are grateful to WATCOM SYSTEMS and WATCOM PUBLICATIONS for their permission to reproduce the reference materials provided in the appendices of this book. These are the reference cards for WATCOM FORTRAN which are`reproduced by courtesy of WATCOM SYSTEMS, and Chapter Three of the USER'S GUIDE for WATCOM FORTRAN authored by D. D. Cowan, P. H. Dirksen, and J. W. Welch, which is reproduced courtesy of WATCOM PUBLICATIONS.

WATCOM FORTRAN is available singly, or by site license, from WATCOM Products Inc., 415 Phillip Street, Waterloo, Ontario, Canada. Tel. 515-886-3700.

Chapter 1
AN OVERVIEW OF WATCOM FORTRAN

The FORTRAN that you will be using is the WATCOM interpreter FORTRAN. There are many different FORTRANs even though it is one of the few languages for which there is a national standard, ANSI 77. Most FORTRANs are different because they are supersets of ANSI 77, that is, they contain more features than the standard. WATCOM contains some features not contained in ANSI 77, but it also lacks a few things that the standard has. In general, these differences concern features that are not often used and are irrelevant to the fundamentals that we will be studying. Programs written in WATCOM have been successfully run with only minor editing on mainframe computers. However, there are three major differences that you need to be aware of.

1. Executable Statements

in standard FORTRAN	must start in a column greater than or equal to column 7
in WATCOM FORTRAN	may start in any column (even columns 1 - 6)

   NOTE: *We strongly recommend that you start in column 7 even when using WATCOM FORTRAN.*

2. Comments

in standard FORTRAN	are denoted by a C in column 1
in WATCOM FORTRAN	are denoted by a * in column 1

3. Continuation Symbols

in standard FORTRAN	are denoted by any symbol in column 6
in WATCOM FORTRAN	are denoted by a & in column 1

WATCOM FORTRAN has four features which make it particularly nice for learning FORTRAN.

---

1. It is an interpretive FORTRAN, not a compiled FORTRAN. Most FORTRANS are *compiled*; this means that after writing a program you must compile it before running it. If you have errors, which is the usual case, you correct the errors and then compile the whole program again before running it, and so on. Compiling a program means translating it into machine language. All high-level languages like FORTRAN have to be translated into machine language sooner or later because the computer works in machine language. In contrast, WATCOM is an *interpretive* FORTRAN which you do not have to compile. The executable statements are translated on the fly as the program is running. This increases the execution time, but it is a sacrifice worth making for learning and program development purposes.

2. WATCOM has an excellent full-screen editor.

3. WATCOM comes with a built-in graphics library. This is very unusual for any FORTRAN. It makes graphics applications much easier than linking to a separate graphics software package (which is the case for other FORTRANs).

4. WATCOM will run on a microcomputer (i.e., any IBM-PC compatible computer). This is typically a friendlier environment than a mainframe computer.

Before you begin programming, you must understand the different types of variables allowed in FORTRAN and the difference between implicit and explicit declaration.

Variable Types

In standard FORTRAN (ANSI 77), a variable name can be one of four simple variable types:

INTEGER	e.g.,	4, 1432, -938, -32768, 0
REAL	e.g.,	4.0, 7.35231, -0.3210, 0.0, 8., -4.
CHARACTER	e.g.,	"bFkd", "-4", "*$xx%", "funny", "R2D2", "87592"
LOGICAL	e.g.,	.true. , .false.

NOTES: *(1) For the CHARACTER variable type, even a number can be considered a character string if it is enclosed in quotes.*
(2) The LOGICAL variable type is not available on WATCOM FORTRAN

Variable Names

A variable name is used to store a value. In WATCOM FORTRAN (*not* standard FORTRAN), a legal variable name conforms to the following rules:

(1) The variable name may contain any number of symbols.

(2) The first symbol of the variable name must be an alphabetic character (i.e., a - z or A - Z).

(3) All other symbols in the variable name may be one of the following: alphabetic character (a-z, A-Z), digit (0-9), underscore character (_), or dollar-sign character ($).

Legal Variable Names	Illegal Variable Names
count	57count
Mother	_Mother
RunNER2	$junk
count456	999
Net_Pay	Net Pay
MONEY$	$MONEY

For variable names, upper and lower case letters are distinguished as *different* letters. Thus, the variable *MONEY$* and the variable *Money$* would be treated as different variables.

(4) Variable names may not be any of the reserved FORTRAN keywords such as PRINT, STOP, END, CALL, and so forth. See Appendix D for a complete list.

Variable Declarations

FORTRAN provides two ways of declaring variables (i.e., bringing a variable into existence): *explicit declaration* and *implicit declaration.*

(1) EXPLICIT DECLARATION

- List the variable names *after* each variable type at the beginning of the program.

```
*   Example of INTEGER, REAL, and CHARACTER declaration statements
*
        INTEGER  A, i, Pay, total
        REAL     number, Sum, INDEX
        CHARACTER*4   name
        CHARACTER     name*4
```

Note: From the above example, you can see that there are two ways to declare a character variable. The first method, character*4 name, is used in either WATCOM or standard FORTRAN. The second method, character name*4, can only be used in WATCOM FORTRAN. Both methods declare a character variable which can take up to four characters (e.g., "test" or "1234").

(2) IMPLICIT DECLARATION (Default Declaration)

- In the absence of *explicit declarations*, if the first letter of a variable is { i, j, k, l, m, n, I, J, K, L, M, N }, then the variable is assumed to be of type INTEGER. Note that the first two letters of the word ***IN***TEGER define the range of the letters mentioned above.
- If the first letter in the variable is any one of the remaining letters, the variable is assumed to be of type REAL.

Warning: Implicit declaration does not cover character variables.

Examples:

Index number jars	} integer	sum Angle total	} real

Remember the following rule whenever you attempt to determine the variable type of a variable name:

EXPLICIT DECLARATIONS ALWAYS OVERRIDE IMPLICIT DECLARATIONS

Chapter 2
Getting Started

Literacy Sequence

What follows is a literacy sequence designed to give you a quick introduction to some of the most fundamental skills. After this, we will start working through the material more methodically.

WATCOM FORTRAN is usually booted (or started) by entering WFOR after the DOS prompt, e.g., A:>WFOR. If you wish to utilize the graphics option, you need to configure the software for graphics by running the CONFIG file. This only has to be done once for any given copy of WATCOM. When you run the CONFIG file, just confirm everything, but choose graphics and the "options" category.

When you boot WATCOM FORTRAN you will be in the editor. This is where you will write your programs. The screen is divided into three different parts: (1) the workspace area or text area, (2) the command line, and (3) the function-key line. Whenever you enter WATCOM FORTRAN from the main menu, the cursor (an underscore character) will be on the *command line*. Above the command line is the *workspace area*; this is where you will type in your programs. Below the command line is the *function-key line* :

F1: Pg↑ 2: Pg↓ 3: L↑ 4: L↓ 5: LIns 6: LDel 7: Sel/Desel 8: Cut 9: Scrn/Cmd 10: Help

For a very brief description of the function keys, press **F10** (one of the ten function keys on the left-hand-side of the keyboard). For a full account of the use of the function keys, see APPENDIX C: WATCOM Reference Card Editor which contains a detailed description of the function keys and other features of the WATCOM FORTRAN editor. *Note: It is to your advantage to learn how the function keys work because they make file editing easier.*

To enter the *workspace area*, press **F9** once. The cursor, the underscore character, should now be on the **<beginning of file>** line in the workspace area (Note: The cursor should "look like" it is sitting between the <beginning of file> line and the <end of file> line.).

To open up a space for writing a program in, press the **F5** key four or five times. To delete any unwanted lines, use the **F6** key.

To move around the workspace area, use the arrow keys on the right-hand-side of the keyboard or **F3** (move up one line) or **F4** (move down one line).

The Assignment Statement

A FORTRAN program is a set of statements which are executed sequentially unless otherwise specified. The most commonly used statement in a FORTRAN program is the *assignment statement* whereby the value of a variable is created or modified in some way. The *assignment operator* is =. The following program shows how the assignment statement is used. Type the program in. *Remember to type your program in column 7 or greater.* You may use either upper or lower case letters as long as you are consistent. For variable names, upper case and lower case letters are treated DIFFERENTLY. For example, *x* and *X* are two different variables. For FORTRAN keywords/terms, upper and lower case are *not* treated differently (i.e., print and PRINT are treated as the same).

```
X=5.0
Y=X+4.
PRINT, X, Y
STOP
END
```

The first two statements are assignment statements. The first assignment statement, *X = 5.0*, assigns the value of 5.0 to the variable X. The second assignment statement, *Y = X + 4.*, assigns the value of 9.0 to the variable Y, since X+4. = 5.0+4.0 = 9.0. The third statement is an output statement which prints out the values of X and Y. And, the fourth and fifth statements are control statements which tell the computer that this is the end of the program.

Press **F9** to return to the command line.

Type 3 on the command line and press ↵ (the ENTER/RETURN key). This makes line 3 of the program (PRINT, X, Y) the current line. If the brightness and contrast controls for the screen are set properly, the current line should be brighter than the others. Press **F9**, and you will enter the workspace area of the screen at line 3. Press **F9** to return to the command line.

Now, type RUN on the command line, and press the ENTER key.

If you receive the following error message, [DEBUG?], then type in **Q**, and press the ENTER key to return to the *WATCOM FORTRAN editor*. Check to make sure that your program is exactly the same as the above program, then try running the program again.

Now, type ***D** on the command line to clear your workspace area, and then type in the following program.

```
X=15
Y = X/4
PRINT Y
STOP
END
```

Press **F9** to return to the command line.

Now, type RUN and press the ENTER key.

Too bad ... your first error message. The PRINT command needs a comma after it. The problem is easy to fix, but first let's look at the debugger. Usually when a program crashes (as it has for this program), control will automatically be transferred to the debugger. There are only two main options in the WATCOM debugger: **E** for execute or **Q** for quit. For this program, try the following,

[DEBUG?]	*computer prompt*
E PRINT, X	*your input*
[15.000000]	*computer output*
E PRINT, Y	*your input*
[3.750000]	*computer output*
Q	*your input*

The option **E** must be used with a WATCOM FORTRAN command. It is very useful for finding out what some of the key variable values are at the time of a crash. This is done by using it with PRINT as we have here. Hence the value of X is printed (15.000000).

The option **Q** is the most used command because it gets you back to the program listing. You will return to the *WATCOM FORTRAN editor.*

Press the **F9** key to enter the workspace area of your program. Note that the line with the error is automatically made the current line.

Now, use the arrow keys to place the cursor below the space after the T in the line PRINT Y. Press the **INSERT** key. This puts you in insert mode - note that the cursor has changed from an underscore character to a block character (meaning insert mode). Type in a comma and then press the **INSERT** key again to leave the insert mode. The cursor has now changed back to the underscore character.

The line should now read PRINT, Y which is correct. Press **F9** to return to the command line and enter RUN. The program should now run properly. After running the program and examining the output, press the ENTER key to return to the *WATCOM FORTRAN editor.*

Editing

If we had not used the insert mode when we edited, we would have been in the replace mode; and the character we entered would have replaced the one already there. In our case, we would have replaced the space with the comma, but we needed the space so we used the insert mode. Note again that the cursor becomes bigger (i.e., a block character) while in the insert mode.

If the character to be edited is both wrong and extra, you may use the **DELETE** key to remove it. You may also remove a mistake with the backspace key .

Integer vs. Real

Edit your program to read as follows:

```
I = 15
J = I/4.0
PRINT, J
STOP
END
```

Press **F9** to return to the command line and type RUN. The output is 3! Why? Because of the implicit declaration rules, the variables I and J are of type integer. Thus, when the second line of the program is executed, the computer calculates that 15/4.0 = 3.75; but since J is an integer variable, only the integer part of 3.75 (i.e., 3) is assigned to J.

Like the first example program you typed in, the above example contains two of the most important characteristics of all computer programs.

(1) A program is essentially a list of statements that are executed sequentially (Note: This is a simplified view of what a program really is.).

(2) A program statement can only be executed if the information required (by that statement) has already been provided. For example, the second line of our program is an assignment statement (=) where the value I/4.0 is assigned to the variable J. However, if I is undefined (i.e., I has not been assigned a value), then J cannot be determined, and the program will crash (i.e., produce an error).

Delete the first line of the program. Return to the command line and type RUN. Read the error message. Any time you get this message, check to see that everything on the right hand side of the assignment statement is defined.

Looping

A very common structure in programming is the *loop*. We will study it extensively in Chapter 4, but we may as well begin the familiarization process and take a look at it now.

You can delete the present program by going to the command line and typing ***D** and pressing ENTER. After doing this, type in the following program,

```
             *     Any line with an asterisk in column 1 is a comment statement.
             *     Comment statements are NOT executed by the computer.
             *
             *     The following program illustrates the loop structure.
                   N = 20
begin loop →       DO 10 I = 1,12
|  contents           PRINT, "I = ", I
|  of the             N = N + 5
|   loop              PRINT, "N = ", N
end loop →    10   CONTINUE
                   STOP
                   END
```

Note that a statement label uses columns 1-5. Also note the practice of indenting the contents of the loop for readability.

Return to the command line and enter RUN to execute the program. After viewing the output, return to the *WATCOM FORTRAN editor* by pressing the ENTER key.

Examine the program. Note the loop syntax. A statement label is used to specify a control sequence. That is, it is only used when the sequential execution of statements needs modifying. For example, the statement – 10 CONTINUE – means to continue execution at the statement labeled with the number 10; that is, execution is transferred to the statement – DO 10 I = 1,12.

The DO/CONTINUE loop contains three major pieces of information which are necessary to control the number of times a loop is executed:

DO s <index> = <initial value>, <test value>, <increment>

(1) a statement label, s, which pairs up the DO statement and the CONTINUE statement.
(2) the name of the loop counter which is called the ***index*** of the D O loop.
(3) the parameters of the DO
  (a) <u>initial value</u> – the starting value for the counter/index
  (b) <u>test value</u> – the final value for the counter/index
  – the value to which the counter is compared after the addition of the increment (see below)
  (c) <u>increment</u> – the value added to the counter *after* each repetition of the loop (if an increment is not specified, it is assumed to be an increment of 1)

One of the most common inefficiencies in constructing loops is in using more variables than necessary. For example, you can modify this program so that is uses only one variable "I" or "N", but still produces the same output. Press **F 9** to enter the workspace area of your program and edit the program as follows:

(1) change *DO 10 I = 1,12* to read *DO 10 I = 25,80,5*
(2) delete the following lines – *N = N + 5* and *PRINT, "N=",N*

Return to the command line. Run the program to see the effect of the new looping specifications.

File Management

The following commands for file management of FORTRAN programs must be entered on the command line:

to save a file	**PUT A:<filename>.FOR**
to retrieve a file	**GET A:<filename>.FOR**
to erase a file	**ERASE A:<filename>.FOR**
to see a list of your files	**DIR A:**
to clear the workspace area	***D**

where <filename> should consist of letters or numbers and is less than or equal to 8 in length.

With the program you have, we can practice these commands.

Go to the command line (if you are not already there) and make sure that your floppy disk is in the disk drive and the lever is down. Enter the following:

PUT A:TEST.FOR

This has saved your file to the floppy disk in drive A (always specified as A:). As you will see it was necessary to add the filetype extension .FOR to identify it as a FORTRAN file.

Next, enter DIR A:

This provides a directory of the files on your floppy disk in drive A. On the screen, you will see your saved file TEST.FOR. To return to the WATCOM editor, press the ENTER key.

Now, delete your program by entering ***D** in the command line. Your program should be gone now.

To retrieve your program, enter the following

GET A:TEST.FOR

Your program is back.

If you wish to delete a program from your directory (i.e. from the floppy disk), use the ERASE command. For example,

ERASE A:TEST.FOR

will erase the program which was saved as TEST.FOR.

If you have use of a hard disk (C:), repeat the exercise using the hard disk. In general, you may omit the drive specification (A: , C:) if you are storing your files on the same drive from which you are running the WATCOM software.

Type in the following program and run it. It is very similar to our earlier loop program.

```
      X=10.
      DO 20 I=1,5
         X=X+1.
         PRINT, X
   20 CONTINUE
      END
```

The assignment statement, *X = X + 1.*, is enormously useful in engineering applications. (Note, again, that prior to its use in the loop the value of X had to be initialized, that is, given an initial value.) However, the statement, *X = X + 1.*, is mathematically impossible if the = is taken to mean *equal to*. But, in actuality, the = sign operates more like a ← sign. That is, take whatever is on the right hand side, evaluate it, and then assign it to the variable on the left hand side.

We will use this program to demonstrate other management techniques. Change the test value of the loop from 5 to 500. When you run the program now, it will be running for a long time outputting the real values 10 to 510. To stop the output, you should hold down the CONTROL and BREAK keys at the same time. Then, enter Q to return to the editor. (Some software will not recognize CONTROL-BREAK, but your WATCOM FORTRAN does.)

Another useful technique is the use of CONTROL-S to temporarily halt the screen output. Run the program again, press down the CONTROL key, and press the S key once quickly. The screen output will freeze. Press any key and it begins again. Obviously, you must be careful to press the S just once and release, otherwise the screen output will commence again, and stop again, and so on, in rapid succession.

Both CONTROL-BREAK and CONTROL-S are peculiar to MS-DOS operating environments. They usually will not work on other systems and they have nothing to do with FORTRAN, per se.

Arithmetic Operators

You have already used + (plus) and / (divide). Edit the above program by replacing *X = X + 1.* with *X = X * 10.* Run the program, and study the results. Now, replace that same line with *X = X - 10.*, and run the program again.

For exponentiation, edit the same line to read *X = 10. ** I* . Note that there is no longer a self-referenced modification, that is, *X* does not appear on the right hand side of the =. This means that the initialization statement, *X = 10.*, is no longer needed and can be discarded.

To evaluate longer arithmetic expressions, the following rules must be followed:

<u>ORDER OF EVALUATION FOR ARITHMETIC EXPRESSIONS</u>

1. Expressions in parentheses, innermost first.
2. Exponentiation, right to left.
3. Multiplication and division, left to right.
4. Addition and subtraction, left to right.

Type in the following program and see if you can correctly predict the output before you run it.

```
*      Sample program with arithmetic operators
*
        V = 2. + 3. / 6.
        X = (2. + 3.) * 5. ** 2
        Y = 2. + 3. +5. ** 2
        Z = (2. + 3. * 5.) ** 2
        PRINT, V, X, Y, Z
        END
```

CHAPTER 2 PROBLEM ASSIGNMENTS

1. Assume your weight increases by 3 lbs. a year for 12 years. Write a program that will print out what your weight will be for each year. Use your present weight as a starting value.

2. Repeat question one, but this time assume your weight decreases by 2 lbs. each year.

3. Repeat question one, but this time assume your weight increases by 3% each year.

Chapter 3
Integer/Real Arithmetic, Functions, and Variable Types

Integer/Real Arithmetic

In the last chapter we studied arithmetic operators, now we will look at some of the special problems which occur when you use integer/real arithmetic.

Integer/real arithmetic can be explained with the following table. Operand 1 is on the vertical axis, and operand 2 is on the horizontal axis.

operand 2

	+, -, *, /, or **	integer	real
operand 1	integer	integer	real
	real	real	real

For example, an integer times an integer results in an integer; but a real number times an integer results in a real number. Thus,

```
3. + 4 = 7.000000     (real + integer = real)
3 * 4 = 12            (integer * integer = integer)
4 / 8 = 0             (integer / integer = integer)
4 / 8. = 0.500000     (integer / real = real)
4. / 8. = 0.500000    (real / real = real)
```

Let's run through a more thorough example:

Problem: Convert 1580 yards to kilometers
Approach used to solve the problem:
- (a) convert yards to miles (1760 yards = 1 mile)
- (b) then convert miles to kilometers (1.61 km = 1 mile)

(1) *Incorrect way for converting 1580 yards to kilometers*

Assuming implicit declaration rules, let x represent kilometers

```
x = (1580/1760) * 1.6
        |          |
yards to miles  miles to kilometers
```

If the command, **print, x**, is executed, the output will be 0.000000

Why?

The reason is integer arithmetic. An integer divided by an integer results in an integer. Since 1580 divides into 1760 **zero** times with a remainder of 180, the result for (1580/1760) is the integer zero! Now, an integer times a real (i.e. 0 * 1.6) results in a real number -- 0.000000, and since it is assigned to a real variable x, the command, **print, x**, will output 0.000000.

What would happen if we used an integer variable k instead of x?
Answer: The output for **print, k** would be 0. WHY?

(2) *Correct ways for converting 1580 yards to kilometers*

(a)	x = (1580/1760.) * 1.6	(Note: decimal point after 1760)
(b)	x = (1580./1760) * 1.6	(Note: decimal point after 1580)
(c)	x = (1580./1760.) * 1.6	(Note: decimal points after both)
(d)	x = (1580 * 1.6) / 1760	Why does this work?

Intrinsic Functions

Since many algorithms are used so repeatedly in FORTRAN programs, these algorithms are often *built-in* the language. Different implementations of FORTRAN vary in which functions are built-in, but all include those most commonly used. For example, taking the square root of a value can be achieved as follows:

```
PRINT, "ENTER A POSITIVE NUMBER"
READ, X
Y = SQRT(X)
PRINT, "THE SQUARE ROOT OF", X, "IS", Y
END
```

Type the program in and run it.

The above program introduces the *READ* statement; this statement tells the computer to receive input from the keyboard. Whatever value is input from the keyboard is assigned to the variable X. Notice that the READ statement requires a comma after it. The first line of the program, *PRINT, "ENTER A POSITIVE NUMBER"*, is not required, but it provides a useful prompt for the user of the program.

SQRT is an intrinsic function because it is built-into the language. It computes the square root of the real quantity in parentheses. Later you will learn how to write your own functions (i.e., extrinsic functions).

All of the available intrinsic functions are listed in Appendix D. Some of the most commonly used are the trigonometric functions such as SIN( ), COS( ), and TAN( ). Note that all the trigonometric functions take real numbers in *radians, not in degrees*. So, if you ask for COS(30), you will get the the cosine of 30 radians, not the cosine of 30 degrees.

Two other functions which are often used are INT( ), which converts a real number to an integer number by truncating all significant digits to the right of the decimal point, and FLOAT( ), which converts an integer number to a real number. If you wish to have these functions demonstrated, type in the following program and run it.

```
X = 31.72
M = INT(X)
Y= FLOAT(M)
PRINT, X, M, Y
END
```

Occasionally, a function will need more than one argument, still others require no arguments at all. However, all functions, both intrinsic and extrinsic, *return* only one value.

Thus far, we have studied two of the four simple variable types: integer and real. We will now study character variables – variables which store character strings. However, before we begin, we will review the use of declaration statements with integer and real variables.

Declaration Statements

As previously described, any variable name beginning with one of the letters I - N (upper or lower case), inclusive, is treated as an INTEGER variable *unless otherwise declared.* Any variable name beginning with any other letter is treated as a REAL variable *unless otherwise declared.* This is the *implicit declaration* situation. To otherwise declare a variable requires explicit declaration(s) at the beginning of the program.

Type in the following program.

```
REAL FAHR
INTEGER CELSIUS
PRINT, "Enter a Fahrenheit temperature"
READ, FAHR
CELSIUS = (FAHR - 32) / 1.8
PRINT, "Celsius temperature = ", CELSIUS
END
```

If you run this program, you will find that the result is somewhat unsatisfactory. The variable CELSIUS is operating as an integer, just as you declared it. However, this introduces a truncation that can result in an error of almost 1 degree Celsius. For example, if you run the program and enter a value of 33.7 degrees Fahrenheit, you will get an answer of 0 degrees Celsius whereas the correct answer is closer to 1 degree Celsius.

To correct this problem, eliminate the declaration statement *INTEGER CELSIUS.* The variable CELSIUS will now revert to its implicit declaration status as a real variable. Now, run the program again, and you will find that it produces the answer as a real, and more accurate, number.

Finally, note that the declaration statement *REAL FAHR* is not really needed since FAHR would still be a real number under the implicit declaration rules. However, good programming practice dictates declaring all variables explicitly. So, to follow good programming practice, we should declare CELSIUS explicitly also.

Character Strings

Since the implicit declaration rules only cover integer and real numbers, characters strings must always be *explicitly* declared. A *character string* is a sequence of characters like a word or even a social security number; it has a symbolic value rather than a numeric value. When a program is executed, character strings are not used in any algebraic fashion; however they are used in assignment statements, and they may be combined (i.e., concatenated) which is a form of addition. It is also possible to determine the length of a character string and to access particular characters from the character string.

Type in the following program.

```
* Program using the variable type CHARACTER
*
        CHARACTER*20    TWO, PROMPT
        CHARACTER       AGE*2, ONE*28, NAME*8
        PROMPT = 'IS YOUR FIRST NAME'
        PRINT, 'ENTER YOUR FIRST NAME'
        READ, NAME
        PRINT, 'ENTER YOUR AGE AS A 2-DIGIT NUMBER'
        READ, AGE
        ONE = NAME // PROMPT
        TWO = 'YOU ARE ' // AGE // ' YEARS OLD'
        PRINT, ONE, ' AND ', TWO
        END
```

From the above program, you should note that there are two ways of declaring a variable to be of type character. The first way is the easiest to use when the character strings are of the same length. For the statement – *CHARACTER*20 TWO , PROMPT* – both TWO and PROMPT are declared to be character strings of length 20. *If no length is specified, then the length of the string defaults to a length of 1.* The second way of declaring a character variable is shown in the statement – *CHARACTER AGE*2, ONE*28, NAME*8* – where the length of each character string is specified after each variable name. Thus, AGE is a character string of length 2; ONE is a character string of length 28; and NAME is a character string of length 8.

The values for the variables ONE and TWO are both formed by concatenation (i.e., the // operator). ONE is formed by using the previously defined strings PROMPT and NAME. TWO is formed by simultaneously creating strings and adding them to a previously defined string, AGE.

The main problem with the above program is the relatively uncontrolled length of the string NAME. *If you do not use the entire length of the string, then the extra characters will be printed out as blank spaces.* For example, if your name is Cheryl, then the value of the variable NAME would be printed out as follows:

Cheryl⊬⊬

where ⊬ means a blank space.

On the other hand, *if you try to exceed your allotment, your output is truncated accordingly.* For example, if your name is Stephanie, the the value of the variable NAME would be printed out as follows:

Stephani

So, character string allotment is left justified. Any unused characters are printed out as blank spaces to the right; and it truncates (without crashing the program) at the right when the assigned limit is reached. For now, we will have to live with this problem until we study formatting in Chapter 7.

Character String Manipulation

One useful characteristic of strings is the ease with which the individual characters may be accessed. For example,

```
* WHAT'S IN A NAME?
*
      CHARACTER NAME*7
      NAME = 'BRENDA'
      PRINT, NAME(2:4)            prints characters 2-4
      PRINT, NAME(1:1)            prints character 1
      PRINT, NAME(5:)             prints characters 5-6
      PRINT, NAME(:5)             prints characters 1-5
      PRINT, NAME                 prints the whole name
      END
```

Run this program and study the output. It should give you a clear idea about accessing the contents of a string.

CHAPTER THREE PROBLEM ASSIGNMENTS

1. Write a program which reads in any angle between 0 and 90 degrees and then produces the cosine, sine, cosine squared, sine squared, and sum of squares (e.g. if x is the angle, then produce cos x, sin x, $\cos^2 x$, $\sin^2 x$, and $\cos^2 x + \sin^2 x$). Use a character string to identify the output like the one that is used in the SQRT program on page 14. Your angle will need to be converted to radian measure by dividing by 180 and multiplying by 3.14159 (pi).

   Note: If you get unexpected answers, read pages 13-14 on integer/real arithmetic again.

2. Write a program using intrinsic functions that will provide the square root of the absolute value of any number. Print out the results in three ways:
   (a) the exact square-root value
   (b) the value as the largest integer less than the exact square-root value
   (c) the answer in part (b) as a real number.

   For example, if the number 20.0 is input, the results would be

   ```
   4.472135954
       4
   4.000000000
   ```

   Note: The second line of the output is indented; that is, there are 4 extra blanks to the left of the integer 4. Why? The computer chose the output format for you. The computer also chose the output formats for the first and third lines of output (although this may not be as apparent). If you are interested in learning how to override the computer's output format now, instead of later; then after you have finished the problem assignments from this chapter, do chapter 7 before proceeding onto chapter 4.

3. Write a program that will take any name as input and find its length – you will need to look up the intrinsic function LENSTR( ) in Appendix D. Create a DO loop that will print out the name vertically on the screen one character at a time. NOTE! You may control the number of loops with the length of the name.

4. Modify the program in 1 so that it will print out the first letter of the name, then the first two letters, then the first three letters, and so on until the whole name is printed out.

Chapter 4
Relational Operators, Program Control, and Loop Structures

In the last three chapters, we have seen that statements in a program are normally executed sequentially. That is, the statements in a program are normally executed in the order in which they appear in the program. However, in this chapter, we will concentrate on *control structures* which alter the sequential execution of a program.

Relational Operators

Before we learn about these control structures, we must first look at the *relational operators* which are used in these control structures. Relational operators are of two types: arithmetic and logical.

ARITHMETIC RELATIONAL OPERATORS

.GT.	$>$	greater than
.GE.	$\geq$	greater than or equal to
.LT.	$<$	less than
.LE.	$\leq$	less than or equal to
.NE.	$\neq$	not equal to
.EQ.	$=$	equal to

LOGICAL RELATIONAL OPERATORS

.AND.	logical AND
.OR.	logical OR
.NOT.	logical NOT

To illustrate the use of these relational operators, six examples are presented.

Assume the following: $J = 4$ and $K = 12$

Then,

(1) (J .LE. 4) → (4 .LE. 4) → $(4 \leq 4)$ → *TRUE*

(2) (J .GT. 8) → (4 .GT. 8) → $(4 > 8)$ → *FALSE*

(3) (K .EQ. 12) → (12 .EQ. 12) → $(12 = 12)$ → *TRUE*

(4) ((J .NE. 10) .AND. (K .EQ. 12)) → ((4 .NE. 10) .AND. (12 .EQ.12))
→ $((4 \neq 10)$ .AND. $(12 = 12))$ → (*TRUE* .AND. *TRUE*) → *TRUE*

(5) ((K .NE. 12) .OR. (J .LT. 8)) → ((12 .NE. 12) .OR. (4 .LT. 8))
→ $((12 \neq 12)$ .OR. $(4 < 8))$ → (*FALSE* .OR. *TRUE*) → *TRUE*

(6) (.NOT. ((K .NE. 12) .OR. (J .LT. 8))) → (.NOT. ((12 .NE. 12) .OR. (4 .LT. 8)))
→ (.NOT. $((12 \neq 12)$ .OR. $(4 < 8)))$ → (.NOT. (*FALSE* .OR. *TRUE*))
→ (.NOT. *TRUE*) → *FALSE*

IF ... THEN

The first control structure that we will study is the *closed conditional IF-THEN*. The syntax (or form) of the closed conditional IF-THEN is as follows:

IF (<relational-expression>) THEN <then-statement>

where

<relational-expression> is an expression which uses a relational operator(s), and
<then-statement> is any valid FORTRAN statement.

The <then-statement> is executed only if the <relational-expression> is true.

The diagram below shows the program flow of control through an IF-THEN structure. This diagram is called a *flow chart*, and it is widely used in programming to "see" the flow of control through a program.

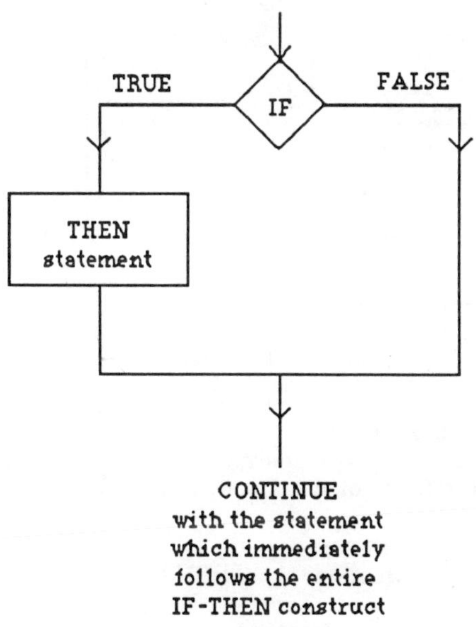

FLOW CHART OF THE CLOSED CONDITIONAL IF-THEN

To illustrate its use, type in the following program and run it.

```
* CLOSED CONDITIONAL IF-THEN EXAMPLE
*
      PRINT, 'ENTER AN INTEGER BETWEEN 0 AND 100'
      READ, K
      IF (K .GE. 90) THEN PRINT, 'YOUR GRADE IS AN A'
      PRINT, 'THE END'
      END
```

If the integer entered is greater than or equal to 90, then
(1) the relational expression (K .GE. 90) is true,
(2) the statement 'YOUR GRADE IS AN A' is printed out,
and (3) the program continues execution at the statement following the IF-THEN construct, namely *PRINT, 'THE END'.*

If the integer entered is *not* greater than or equal to 90, then
(1) the relational expression (K .GE. 90) is false,
and (2) the program continues execution at the statement following the IF-THEN construct, namely *PRINT, 'THE END'.*

The flow chart for the program is shown below:

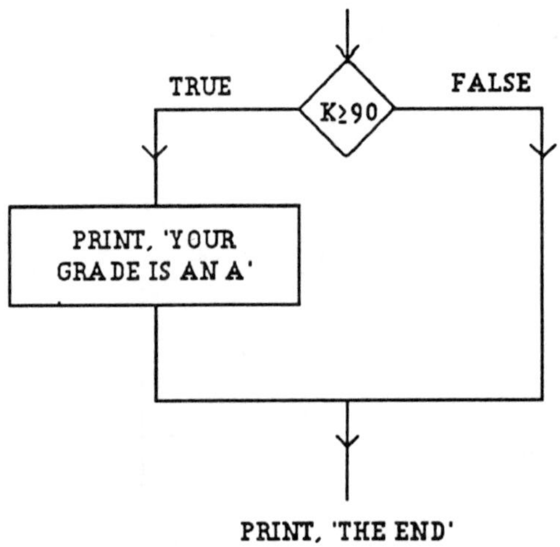

IF ... THEN ... ELSE ... ENDIF

The next control structure that we will study is the *open conditional IF-THEN-ELSE-ENDIF*. The syntax (or form) of the open conditional IF-THEN-ELSE-ENDIF is as follows:

```
IF  ( <relational-expression> )  THEN
      :
      :        <statement(s)>             ]— the THEN block
      :
ELSE
      :
      :        <statement(s)>             ]— the ELSE block
      :
ENDIF
```

where

<relational-expression> is an expression which uses a relational operator(s),
<statement(s)> is one or more valid FORTRAN statements, and
ELSE and the ELSE block are *optional.*

The THEN block is executed if the <relational-expression> is true, otherwise the ELSE block is executed (if there is an ELSE block).

The flow chart for the IF-THEN-ELSE-ENDIF structure is shown below:

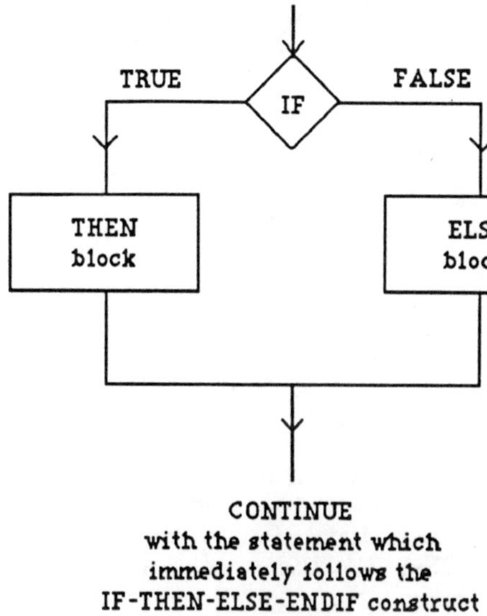

FLOW CHART OF THE OPEN CONDITIONAL IF-THEN-ELSE-ENDIF

The IF-THEN-ELSE-ENDIF control structure creates two alternative program blocks, the THEN block and the ELSE block. Depending on the <relational-expression>, the program will flow either through the THEN block or through the ELSE block, but *NEVER BOTH*.

As previously stated, the ELSE block is optional. When it is *not* included, the syntax and flow chart are as follows:

```
IF  ( <relational-expression> )  THEN
      :
      :        <statement(s)>              ]- the THEN block
      :
ENDIF
```

where

<relational-expression> is an expression which uses a relational operator(s), and
<statement(s)> is one or more valid FORTRAN statements

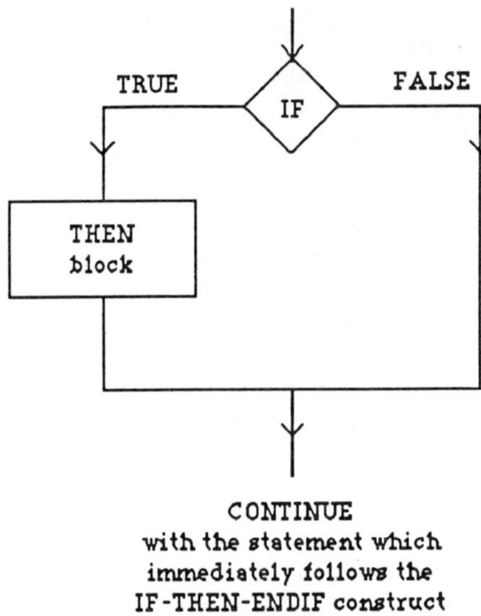

FLOW CHART OF THE OPEN CONDITIONAL IF-THEN-ENDIF

To illustrate the use of the open conditional IF-THEN-ELSE-ENDIF, type in the following program and run it.

```
* OPEN CONDITIONAL IF-THEN-ELSE-ENDIF EXAMPLE
*
      PRINT, 'ENTER AN INTEGER BETWEEN 0 AND 100'
      READ, K
      IF (K .GE. 90) THEN
          PRINT, 'YOUR GRADE IS AN A'
      ELSE
          PRINT, 'YOUR GRADE IS  B, C, D, OR F'
      ENDIF
      PRINT, 'THE END'
      END
```

In this program, there is only one statement in each block; however, the blocks could contain *any* number of statements. Notice that the blocks must begin on the line following THEN or ELSE, and they should be indented for readability.

As before, if the integer entered is greater than or equal to 90, then

(1) the relational expression (K .GE. 90) is true,

(2) the statement 'YOUR GRADE IS AN A' is printed out,

and (3) the program continues execution at the statement following the IF-THEN-ELSE-ENDIF construct, namely *PRINT, 'THE END'.*

However, if the integer entered is *not* greater than or equal to 90, then

(1) the relational expression (K .GE. 90) is false,

(2) the statement 'YOUR GRADE IS B, C, D, OR F' is printed out,

and (3) the program continues execution at the statement following the IF-THEN-ELSE-ENDIF construct, namely *PRINT, 'THE END'.*

The flow chart for the program is shown below:

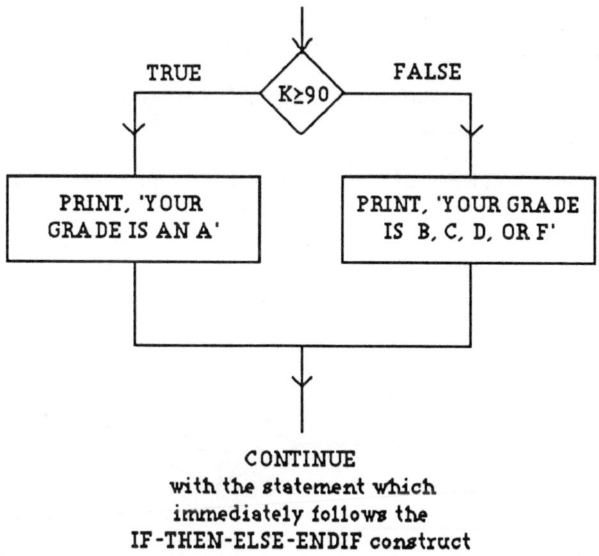

<u>IF ... THEN ... ELSEIF ... ELSE ... ENDIF</u>

With the IF-THEN-ELSE-ENDIF structure, only two alternative flows of control are possible – the THEN block and the ELSE block. However, when *several* alternatives are possible, the ELSEIF statement (which is inserted into the IF-THEN-ELSE-ENDIF structure) can be used to select one of any number of alternatives.

The syntax for the IF-THEN-ELSEIF-ELSE-ENDIF structure is as follows:

```
IF  ( <relational-expression> )  THEN
      :
      :          <statement(s)>                      ]- the THEN block
      :
ELSEIF   ( <relational-expression> ) THEN
      :
      :          <statement(s)>                      ]- ELSEIF block
      :
ELSE
      :
      :          <statement(s)>                      ]- the ELSE block
      :
ENDIF
```

where

<relational-expression> is an expression which uses a relational operator(s),
<statement(s)> is one or more valid FORTRAN statements, and
the ELSEIF block and the ELSE block are *optional.*

Although the above program has only one ELSEIF block, it is important to remember that there can be ***any*** *number of ELSEIF blocks.*

The flow chart for the IF-THEN-ELSEIF-ELSE-ENDIF structure is shown below:

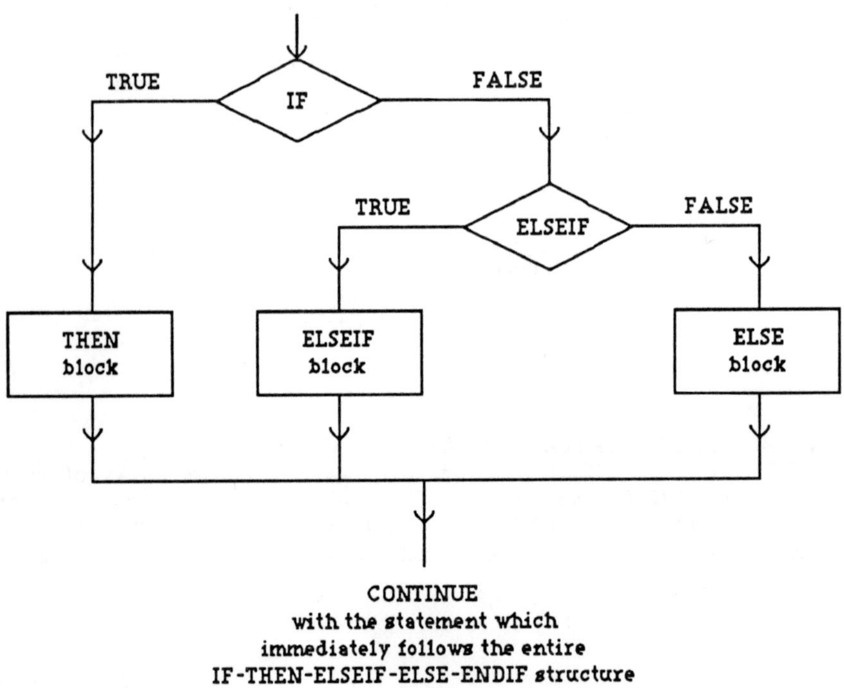

FLOW CHART FOR THE IF-THEN-ELSEIF-ELSE-ENDIF STRUCTURE

The IF-THEN-ELSEIF-ELSE-ENDIF control structure creates two or more alternative program blocks depending on the number of ELSEIF blocks. For example, in the flow chart on the previous page, there are three alternative program blocks – the THEN block, the ELSEIF block, and the ELSE block.

To illustrate the use of the IF-THEN-ELSEIF-ELSE-ENDIF structure, type in the following program and run it.

```
PRINT, 'ENTER AN INTEGER BETWEEN 0 AND 100'
READ, K
IF (K .GE. 90) THEN
     PRINT, 'YOUR GRADE IS AN A'
ELSEIF (K .GE. 80) THEN
     PRINT, 'YOUR GRADE IS A B'
ELSEIF (K. GE. 70) THEN
     PRINT, 'YOUR GRADE IS A C'
ELSEIF (K .GE. 60) THEN
     PRINT, 'YOUR GRADE IS A D'
ELSE
     PRINT, 'YOUR GRADE IS AN F'
     PRINT, 'YOU FAILED THE EXAM'
ENDIF
PRINT, 'THE END'
END
```

In the above program, there are five alternative program blocks -- one THEN block, three ELSEIF blocks, and an ELSE block.

As before, if the integer entered is greater than or equal to 90, then
(1) the relational expression (K .GE. 90) is true,
(2) the statement 'YOUR GRADE IS AN A' is printed out,
and (3) the program continues execution at the statement following the IF-THEN-ELSEIF-ELSE-ENDIF construct, namely *PRINT, 'THE END'*.

If the integer entered is *not* greater than or equal to 90 but is greater than or equal to 80, then
(1) the relational expression (K .GE. 80) is true,
(2) the statement 'YOUR GRADE IS A B' is printed out,
and (3) the program continues execution at the statement following the IF-THEN-ELSEIF-ELSE-ENDIF construct, namely *PRINT, 'THE END'*.

If the integer entered is *not* greater than or equal to 80 but is greater than or equal to 70, then
(1) the relational expression (K .GE. 70) is true,
(2) the statement 'YOUR GRADE IS A C' is printed out,
and (3) the program continues execution at the statement following the IF-THEN-ELSEIF-ELSE-ENDIF construct, namely *PRINT, 'THE END'*.

If the integer entered is *not* greater than or equal to 70 but is greater than or equal to 60, then

(1) the relational expression (K .GE. 60) is true,
(2) the statement 'YOUR GRADE IS A D' is printed out,
and (3) the program continues execution at the statement following the IF-THEN-ELSEIF-ELSE-ENDIF construct, namely *PRINT, 'THE END'*.

If the integer entered is *not* greater than or equal to 60, then

(1) the relational expression (K .GE. 60) is false,
(2) the statements 'YOUR GRADE IS A F' and 'YOU FAILED THE EXAM' are printed out,
and (3) the program continues execution at the statement following the IF-THEN-ELSEIF-ELSE-ENDIF construct, namely *PRINT, 'THE END'*.

The flow chart for the program is shown below:

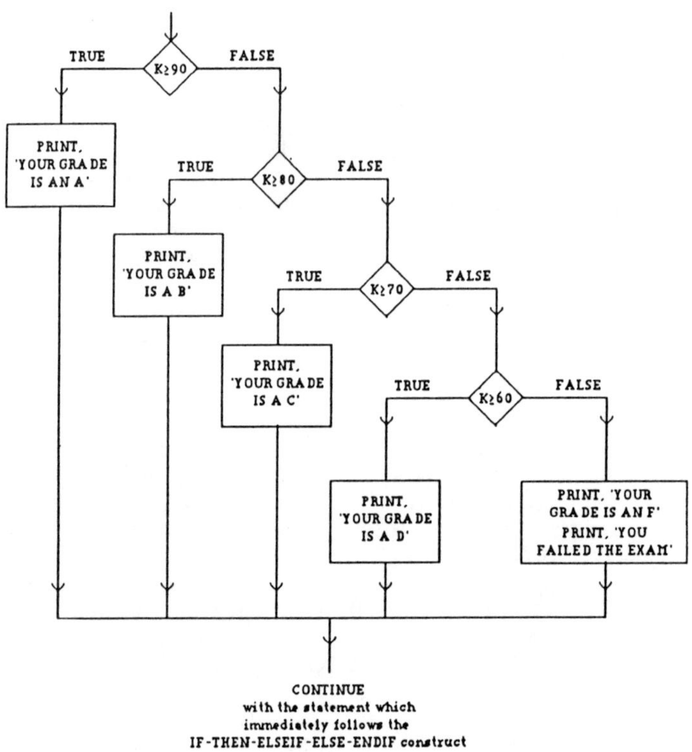

Loops, Loops, and More Loops

In addition to the standard DO/CONTINUE loop construct, WATCOM has a number of others. Type in the following program and run it as one program. The * (comment) lines separate the loops. Note the indentation for the content of the loops. This makes the listing easier to read and diagnose.

```
* DO/CONTINUE IS THE STANDARD LOOP SYNTAX
* FOR ANSI 77 FORTRAN (STANDARD FORTRAN).
*
      DO 10 I = 1,8
         M = I * I
         PRINT, I, M
   10 CONTINUE
*
*
* DO/ENDDO IS NOT PART OF STANDARD FORTRAN,
* BUT IT IS VERY CONVENIENT AND AVAILABLE IN
* WATCOM.
*
      DO I = 1,8
         M = I * I
         PRINT, I, M
      ENDDO
*
*
* THE WHILE/ENDWHILE IS NOT PART OF STANDARD
* FORTRAN, BUT IT IS AVAILABLE IN ALMOST ALL
* FORTRANS. IT IS THE DE FACTO STANDARD FOR A
* CONDITIONAL LOOP STRUCTURE.
*
      I = 1
      WHILE (I .LT. 9)
         M = I * I
         PRINT, I, M
         I = I + 1
      ENDWHILE
*
*
* THE LOOP/UNTIL STRUCTURE IS NOT PART OF
* STANDARD FORTRAN, BUT IT IS A CLEAR WAY OF
* ESTABLISHING A CONDITIONAL LOOP WHICH
* TERMINATES ONLY ON A CONDITION BEING MET.
*
      I = 1
      LOOP
         M = I * I
         PRINT, I, M
         I = I + 1
      UNTIL I = 9
*
      END
```

As a review, the *DO/CONTINUE loop* syntax is shown below. Notice that it contains three major pieces of information which are necessary to control the number of times a loop is executed.

```
      DO  s  <index> = <initial value>, <test value>, <increment>
                :
                :
                :
   s  CONTINUE
```

(1) a statement label, s, which pairs up the DO statement and the CONTINUE statement.
(2) the name of the loop counter which is called the *index* of the DO loop.
(3) the parameters of the DO
   (a) <u>initial value</u> – the starting value for the counter/index
   (b) <u>test value</u> – the final value for the counter/index
   – the value to which the counter is compared after the addition of the increment (see below)
   (c) <u>increment</u> – the value added to the counter *after* each repetition of the loop (if an increment is not specified, it is assumed to be an increment of 1)

Notes:
(1) If an increment is not specified, it is assumed to be an increment of 1.
(2) "Negative increments" are allowed in WATCOM and ANSI 77 (e.g., DO 10 I = 8, 1, -1).

The *DO/ENDDO loop* is identical to the DO/CONTINUE loop except that no statement label, s, is needed to pair up the DO statement and the ENDDO statement. This is the syntax of a DO/ENDDO loop.

```
      DO  <index> = <initial value>, <test value>, <increment>
                :
                :
                :
      ENDDO
```

From now on, we will refer to both the DO/CONTINUE loop and the DO/ENDDO loop as being DO loops.

The *WHILE/ENDWHILE loop* is a conditional loop; that is, the loop is executed only if the WHILE condition is true. The syntax for a WHILE/ENDWHILE loop is shown below.

```
WHILE ( <condition> )
        :
        :
        :
ENDWHILE
```

where <condition> is any valid relational expression.

In the sample program on page 28, notice that the WHILE/ENDWHILE loop has an extra statement which the DO loops do *not* have. This extra statement, *I = I + 1*, is needed to explicitly modify the value of the counter variable (i.e., I) so that it does not get into an infinite loop (i.e., a loop which executes an infinite number of times). From this discussion, two questions should arise -- (1) Why don't the DO loops need this extra statement? and (2) Why would the WHILE/ENDWHILE loop get into an infinite loop without the extra statement, *I = I + 1*? Let's handle these questions separately.

Why don't the DO loops need the extra statement?

By examining the DO/CONTINUE loop from page 28 and the DO/CONTINUE loop syntax from page 29, we find that this extra statement is included *implicitly* within the DO statement itself.

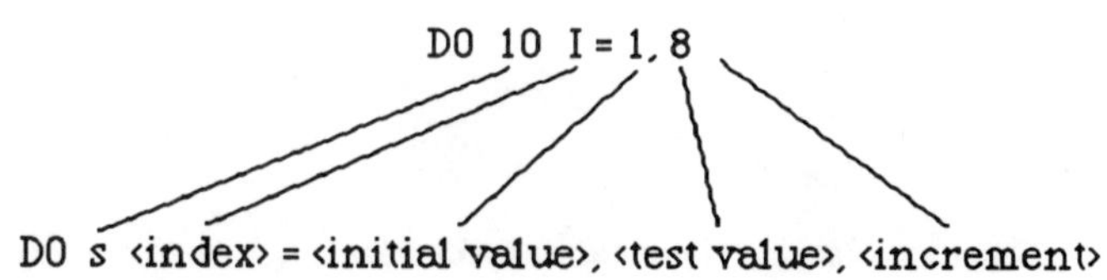

Since no <increment> is explicitly specified, it is assumed to be an increment of 1, that is, I = I + 1. Every time the DO statement is encountered I = I + 1 is executed, and the resulting value is compared to the <test value>.

Why would the WHILE/ENDWHILE loop get into an infinite loop without the extra statement?

By examining the WHILE/ENDWHILE loop without the extra statement, the WHILE/ENDWHILE loop from page 28 would look like this:

```
I = 1                              line 1
WHILE (I .LT. 9)                   line 2
    M = I * I                      line 3
    PRINT, I, M                    line 4
ENDWHILE                           line 5
```

On line 1, I is assigned the value of 1. Line 2 marks the beginning of the first iteration of the loop. Since *(I .LT. 9)* is true, the contents of the loop are executed. On line 3, M = I * I = 1 * 1 = 1. On line 4, I and M are both printed out. When line 5 is executed, control is transferred back to line 2 for a possible second iteration of the loop. (Remember, I still has the value of 1.) Again, since *(I .LT. 9)* is true, the contents of the loop are executed. On line 3, M = I * I = 1 * 1 = 1. On line 4, I and M are both printed out. When line 5 is executed, control is transferred back to line 2 for a possible third iteration of the loop. This process continues on and on and on and on .... Why? Because the variable I will *always* have the value of 1. Since 1 is less than 9, the WHILE condition is *always* true; therefore, the loop is *always* executed over and over again. That is, it is executed an infinite number of times.

In order to correct this problem, a statement which modifies the value of I must be inserted in the loop; and the statement must modify the variable I in such a way so that the WHILE condition becomes false at some future time.

The *LOOP/UNTIL loop* is also a condition loop. The syntax of a LOOP/UNTIL loop is as follows:

```
LOOP
        :
        :
        :
UNTIL ( <condition> )
```

where <condition> is any valid relational expression.

Whereas the WHILE/ENDWHILE loop executes only if <condition> is true, the LOOP/UNTIL loop executes only if <condition> is false. That is, the LOOP/UNTIL loop continues to execute *until* <condition> is true.

In summary, all of the above loops do the same thing, but in some situations the conditional commands WHILE and UNTIL are more appropriate than the DO loops with a fixed number of loops. Notice that the conditional loops have to work with a counter variable (i.e., I) since the loop commands do not contain one.

You should be aware that DO/CONTINUE is standard FORTRAN and that the other possibilities are not always available in a given implementation of FORTRAN although WHILE/ENDWHILE is widely used.

Nested Loops

Loops may be repeatedly nested although two or three levels of nesting are the most often seen. *Nesting means to place loops within other loops.* It is very important that they do not cross each other. Type in and run this program,

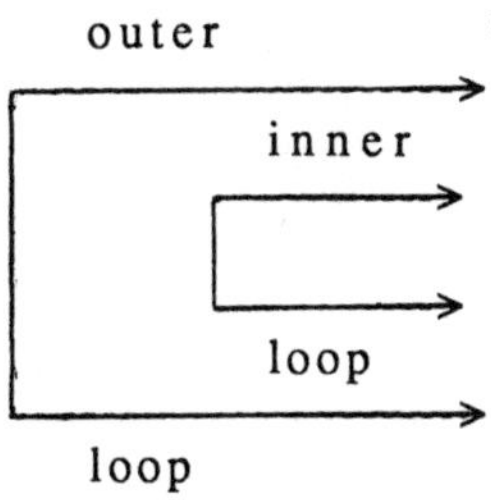


```
* A SIMPLE EXAMPLE OF NESTED LOOPS
*
*
      DO 20 K = 1,2
         PRINT, 'K before inner loop = ', K
         DO 10 J = 1,4
            PRINT, 'J = ' , J
10       CONTINUE
         PRINT, 'K after inner loop = ', K
20    CONTINUE
      END
```

Study the output carefully. Notice that the inner loop must be completed before the outer loop begins another iteration!

Now, let's move on to a more complex example of nested loops.

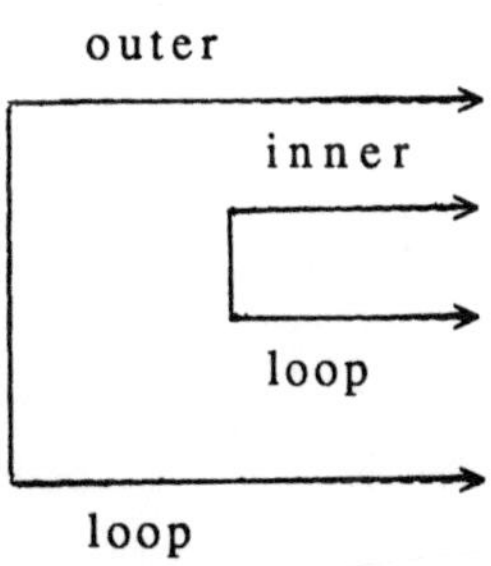


```
* EXAMPLE OF NESTED LOOPS
*
      SUM = 0
      DO 20 K = 1,10
         FAC = 1
         DO 10 J = 1,K
            FAC = FAC * J
10       CONTINUE
         PRINT, K, FAC
         SUM = SUM + FAC
20    CONTINUE
      PRINT, SUM
      END
```

The inner loop in this program calculates the factorial of K where K is controlled by the outer loop. So, the program calculates 1!, then 2!, then 3!, and so on until 10!. In addition, the outer loop keeps a running total for all the factorials added together. The output gives the number and its factorial for each number from 1 to 10 and then the sum of these factorials.

CHAPTER 4 PROBLEM ASSIGNMENTS

1. Using closed conditional IF-THEN statements, write a program that takes a character string input and finds its length. If the length is less than 5, a message "too short" should be printed. If it is longer than 10, a message "too long" should be printed. If the length is between 5 and 10, inclusive, the output should be "just right".

2. Rewrite the program for problem 1 using a IF-THEN-ELSEIF-ELSE-ENDIF structure.

3. Consider the general quadratic $Ax^2 + Bx + C = 0$. If $A \neq 0$, then the roots are

$$x = \frac{-B \pm SQRT(B^2 - 4AC)}{2A}$$

If $A = 0$, then the root is $-C/B$.

Write a program that prints out the solution(s) to the general quadratic equation. Check first if $A = 0$. If $A = 0$, then $x = -C/B$; otherwise check for complex and real roots. If the roots are complex, you may simply print a message saying, "No real roots." However, if the roots are real, then the roots should be printed out to the screen. Use IF/THEN/ELSE/ENDIF structures.

Chapter 5
Arrays and Sorts

ARRAYS

For the last four chapters, we have used three ***simple*** variable types: integer, real, and character. We will now look at a ***structured*** variable type called an array.

An *array* is a structured variable type which is made up of homogeneous, simple variable types.

One-Dimensional Array

A one-dimensional array is simply a list which contains homogeneous data. Example (1) shows a one-dimensional array called WEIGHT containing 7 integers. Example (2) shows a one-dimensional array called SCORE containing 9 real numbers. And, example (3) shows a one-dimensional array called NAME containing 5 strings of length 3.

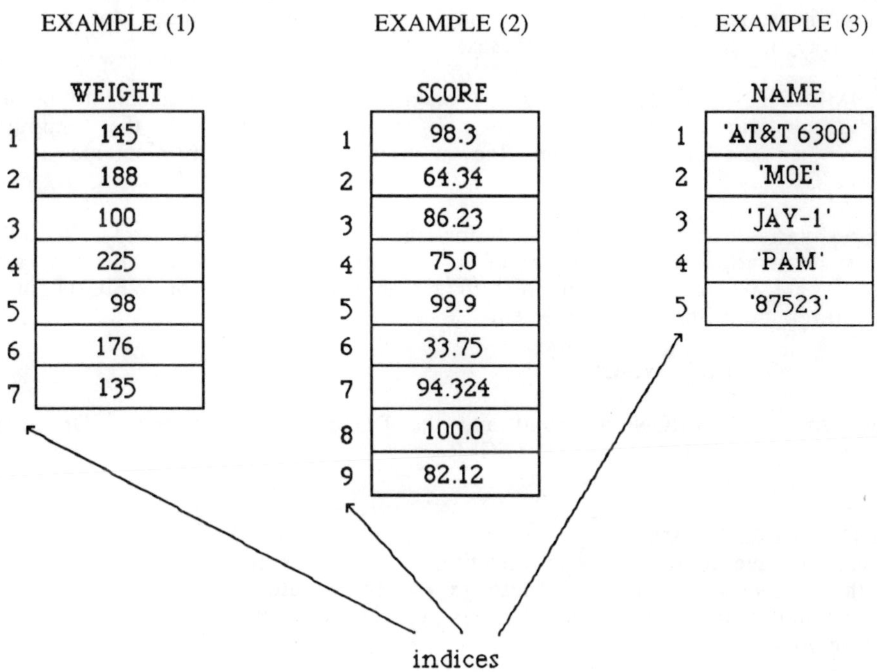

The numbers on the left-hand-side of the arrays are called *indices*. These indices are used to access elements of the array. For example,

WEIGHT(6) contains the value 176.	(The index is 6.)
SCORE(5) contains the value of 99.9.	(The index is 5.)
NAME(1) contains the value of "JOE".	(The index is 1.)

In the above examples, we looked at arrays which already contain data. But how does this data get into the array to begin with? That is, how do we write a program which stores the data into the array (i.e., initializes the array)? The following program shows how to do this. Type it in and run it.

```
* EXAMPLE OF A ONE-DIMENSIONAL ARRAY
*
      DIMENSION  WEIGHT(5)
      DO I = 1,5                                  a loop which initializes
         PRINT, "ENTER WEIGHT NUMBER ", I           the WEIGHT array
         READ, WEIGHT(I)
      ENDDO
      DO I = 5,1,-1                               a loop which prints out
         PRINT, WEIGHT(I)                           the WEIGHT array
      ENDDO                                         in reverse order
      END
```

To run this program to completion, enter 5 numbers representing people's weights; these 5 numbers initialize the array elements in WEIGHT (the 5 numbers are stored in the WEIGHT array). Notice that the index (or subscript) of an array is enclosed in parentheses.

The DIMENSION statement assigns the maximum number of elements to be used in the array and reserves the necessary amount of space in the computer's memory. In the above program, the maximum number of elements in the array WEIGHT is 5.

Since no variable type is specified for the elements in the array WEIGHT, implicit declaration rules are in effect. Thus, the elements in the array WEIGHT are *real* numbers. In order to get an *integer* array of WEIGHT in the above program, insert the statement

```
      INTEGER WEIGHT
```

between the DIMENSION statement and the DO I = 1,5 statement. Or, simply change the word DIMENSION to INTEGER.

Multi-Dimensional Array

Arrays may have more than one dimension. Consider the coordinates of points in a three-dimensional space. These (x,y,z)-coordinates may be contained in three separate one-dimensional arrays or in one two-dimensional array as shown below.

```
* EXAMPLE OF A TWO-DIMENSIONAL ARRAY
*     which contains 7 (x,y,z) coordinates
*
      DIMENSION  IPOS(7,3)
      DO 10 K = 1,7
         IPOS(K,1) = K
         IPOS(K,2) = K*K
         IPOS(K,3) = K+1
         PRINT, (IPOS(K,J), J = 1,3)              implied DO-loop
   10 CONTINUE
      END
```

The two-dimensional array, IPOS, contains 7 rows and 3 columns. The 7 rows indicate the number of three-dimensional coordinates; while the 3 columns indicate the x, y, and z-coordinates respectively. In other words, the first column contains the 7 x-coordinates; the second column contains the 7 y-coordinates, and the third column contains the 7 z-coordinates. The resulting 7*3 matrix is shown on the next page.

1	1	2
2	4	3
3	9	4
4	16	5
5	25	6
6	36	7
7	49	8

WARNING: ALL of the elements in an array must be of the SAME variable type! (In the above example, all the array elements are integers.)

Selecting Elements In An Array

To show how particular elements of an array are selected, type in the following program and run it. The program asks the user to enter 5 names, and then to enter an integer *N* which will select the *N*th name from the NAMES array.

```
CHARACTER*10  NAMES(5)
DO  N = 1,5
   PRINT, 'Enter a name'
   READ, NAMES(N)
ENDDO
PRINT, 'Enter a number between 1 and 5 inclusive'
READ, N
IF ((N .GE. 1) .AND. (N .LE. 5)) THEN
   PRINT, 'The Nth name of the NAMES array is ', NAMES(N)
ELSE
   PRINT, 'You did not enter a number between 1 and 5'
ENDIF
END
```

Finding The Maximum And Minimum Values Of An Array

Finding the maximum and minimum values in an array is easy to do in a loop. The following program shows two different methods for doing this. Type it in and run it.

```
*  METHOD 1:  FINDING MAX AND MIN VALUES
*
        DIMENSION  NRR(5)
        DO  I = 1,5
           PRINT, "ENTER AN INTEGER"              initialize  the
           READ,  NRR(I)                            NRR  array
        ENDDO
        MAX = NRR(1)                         set MAX and MIN equal to
        MIN  =  NRR(1)                       the first element in NRR
        DO  I = 2,5
           IF  (NRR(I)  .GT.  MAX)  THEN  MAX = NRR(I)
           IF  (NRR(I)  .LT.  MIN)  THEN  MIN  =  NRR(I)
        ENDDO
        PRINT,  MAX,  MIN
*
*  METHOD 2:  FINDING MAX AND MIN VALUES
*                     WITH INTRINSIC FUNCTIONS
*
        MA  =  NRR(1)
        MI  =  NRR(1)
        DO  I = 2,5
           MA  =  MAX0(NRR(I),MA)
           MI  =  MIN0(NRR(I),MI)
        ENDDO
        PRINT,  MA,  MI
        END
```

In the first method for finding MIN/MAX values, study the IF/THEN logic, and note the need to set MAX and MIN to the initial value of the array to start the comparisons. In the second method, the intrinsic functions MAX0 and MIN0 do the same thing as the IF/THEN statements in method 1.

To understand selecting better, we will briefly trace through method 1 to see how it works on the following array, NRR.

	NRR	
1	57	NRR(1) = 57
2	38	NRR(2) = 38
3	69	NRR(3) = 69
4	18	NRR(4) = 18
5	90	NRR(5) = 90

initially	I = 2	I = 3	I = 4	I = 5
MAX = 57	MAX = 57	MAX = 69	MAX = 69	MAX = 90
MIN = 57	MIN = 38	MIN = 38	MIN = 18	MIN = 18

SORTING

Selection Sort

Sorting a list of numbers is more difficult than just selecting a particular number from an array. For example, suppose we wish to order a list of numbers in ascending order. First, we would find smallest number; then we would find the smallest of the remaining numbers, and so on until all the numbers are ranked in order. This approach to sorting is called *selection sort.* Selection sorting involves a systematic interchanging of a pair of array elements if they are out of order. Although the interchanging of any pair of values, say X and Y, sounds easy and straightforward, it is actually a little bit tricky. The natural but naive approach might be to

	1.	set X equal to the value of Y,	(X = Y)
and	2.	set Y equal to the value of X.	(Y = X)

However, setting X equal to the value of Y wipes out the previous value of X; this causes both X and Y to have the original value of Y. The second step simply resets Y to its original value. For example,

X = 4, Y = 5 → *set X equal to the value of Y* → X = 5, Y = 5 → *set Y equal to the value of X* → X = 5, Y = 5

Thus, the problem is that we lose the original value of X (i.e., 4) when we set X equal to the value of Y. To remedy this problem, we merely have to save the original value of X temporarily.

To illustrate the problem more clearly, we should consider the interchanging of two values similar to the interchanging of the contents of two cups. Assume that one cup contains tea and that the other cup contains coffee. In order to interchange the contents of the two cups, a third cup is needed to temporarily hold the contents of either the tea or coffee. Let us start by pouring the tea into the third cup. Next, we pour the coffee into the cup that previously contained the tea. Finally, the tea, which is temporarily held in the third cup, is poured into the cup which previously held the coffee. Analogously, the interchanging of a pair of values can be done using the same type of approach but with X = tea cup, Y= coffee cup, and Z = third cup. So, the resulting approach or algorithm for interchanging a pair of values, X and Y, would be as follows:

	1.	save the value of X in Z	*(Z = X)*
	2.	set X equal to the value of Y	*(X = Y)*
and	3.	copy the value of Z into Y	*(Y = Z)*

This algorithm can also be used to interchange a pair of elements in an array. In this case, the approach would be as follows for the array NRR:

	1.	save the value of NRR(J) in NSWAP	*(NSWAP = NRR(J))*
	2.	set NRR(J) equal to the value of NRR(I)	*(NRR(J) = NRR(I))*
and	3.	copy the value of NSWAP into NRR(I)	*(NRR(I) = NSWAP)*

where NRR is an array, NSWAP is a temporary variable used for storage, and I does not equal J.

With the knowledge of swapping values in hand, we present the selection sort as shown below. Pay particular attention to the IF/THEN/ENDIF construct. A pair of array elements will be swapped only when the IF-condition is true. If you do not understand the selection sort, look at the example which follows the FORTRAN program for selection sort, and then come back to look at the program.

```
*   SELECTION SORT (ascending order)
*       assuming that the NRR array is initialized as in
*       the MAX/MIN selection program on page 37
*
        DO J = 1,4
            DO I = J+1,5
                IF (NRR(J) .GT. NRR(I)) THEN
                    NSWAP = NRR(J)
                    NRR(J) = NRR(I)
                    NRR(I) = NSWAP
                ENDIF
            ENDDO
        ENDDO
        END
```

Note: *This particular version of the selection sort can be written more efficiently. Can you see how?*

To illustrate the operation of the selection sort, an example with the array NRR is presented on the following page.

ILLUSTRATION OF THE SELECTION SORT

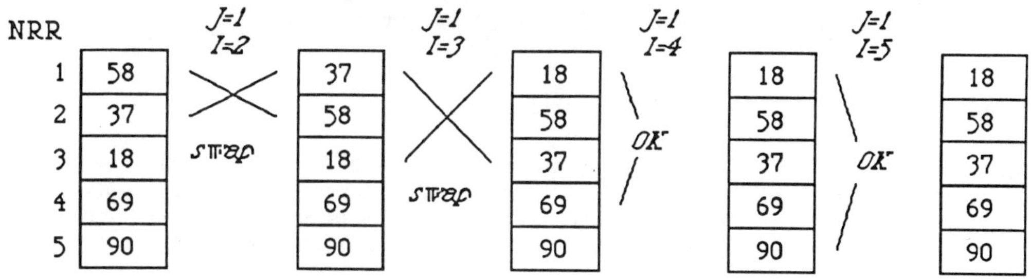

J = 1 is done.

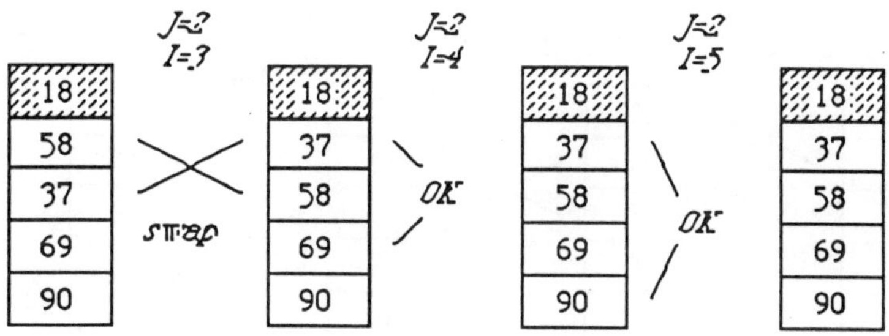

J = 2 is done.

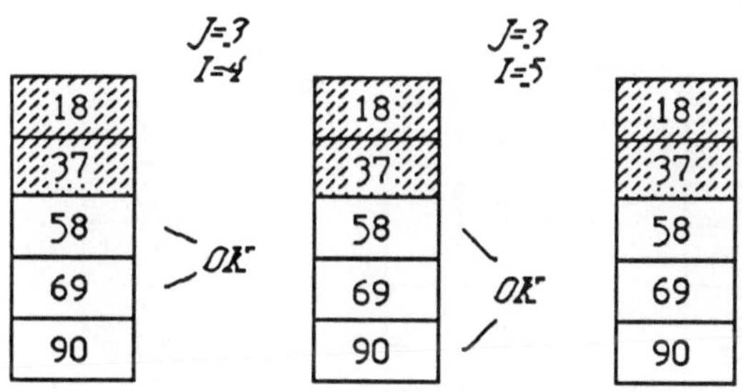

J = 3 is done.

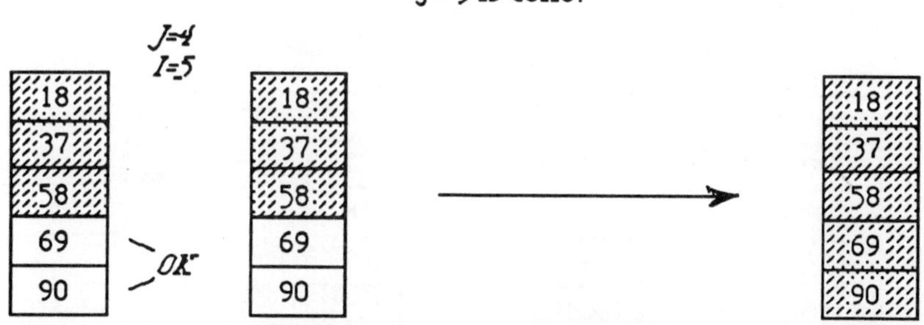

J = 4 is done. ===> SELECTION SORT IS DONE.

Although the selection sort conforms more naturally to the way a human would sort, there are other sorting methods such as the *bubble sort* shown below. In addition, there are many *more efficient* sorting methods (e.g., merge sort, quick sort, and heap sort), but these are considerably more complex and will not be covered in this course.

<u>OPTIONAL: *Bubble Sort*</u>

Bubble sorting involves interchanging pairs of *adjacent* array elements if they are out of order. When every adjacent pair of array elements has been compared, we say that one *pass* has been made through the array. To illustrate the bubble sort, we will use the same array that we used to illustrate the selection sort to sort an array of numbers into ascending order.

ILLUSTRATION OF THE BUBBLE SORT

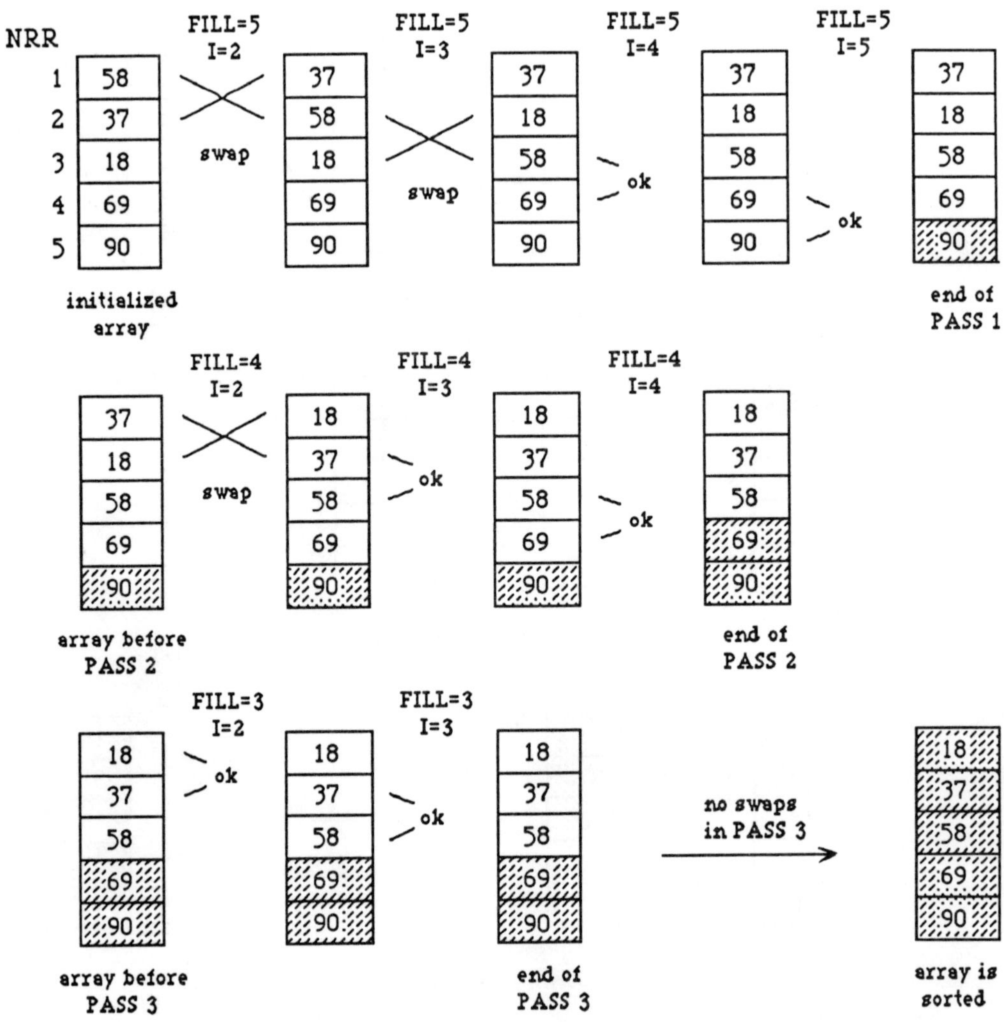

The bubble sort recognizes that sorting is finished when there are no swaps made in a given pass through an array. In the above example, the third pass through the array contains no swaps. So, the algorithm for writing a bubble sort is as follows:

1. set FLAG equal to 1 (indicates that the array is not sorted)
2. while (FLAG = 1) do
   a. set FLAG equal to 0 (assume array is sorted until proven otherwise)
   b. start with the first two array elements
   c. repeat the following until the end of the array is reached
      (1) if current pair of array elements are out of order, then
          (a) swap the array elements
          (b) set FLAG equal to 1 (indicates that a pair of elements have been swapped and that the array is not sorted)
      (2) move to the next pair of array elements

Converting this algorithm into actual WATCOM FORTRAN code, we get the following:

```
*   BUBBLE SORT  (ascending order)
*      assuming that the NRR array is initialized as in
*      the MAX/MIN selection program on page 37
*
       FLAG = 1
       WHILE (FLAG .eq. 1)
          FLAG = 0
          DO I = 2,5
             IF (NRR(I-1) .GT. NRR(I)) THEN
                NSWAP = NRR(I-1)
                NRR(I-1) =  NRR(I)
                NRR(I) = NSWAP
                FLAG = 1
             ENDIF
          ENDDO
       ENDWHILE
       PRINT, (NRR(I), I = 1,5)
       END
```

If you use the bubble sort instead of the selection sort on the NRR array, you should get the same results.

Some of you may have noticed that the *program* for the bubble sort goes through the entire array on *every pass* whereas the *illustration* of the bubble sort does not go through the entire array on every pass. Thus, the illustration of the bubble sort is actually more efficient than the program that we have written; but this problem can be easily fixed by using another variable in the bubble sort program.

```
*    MODIFIED BUBBLE SORT
*        assuming that the NRR array is initialized as in
*        the MAX/MIN selection program on page 37
*
         FLAG = 1
         FILL = 5
         WHILE ((FLAG .eq. 1) .AND. (FILL .ge. 2))
            FLAG = 0
            DO I = 2,FILL
               IF (NRR(I-1) .GT. NRR(I)) THEN
                  NSWAP = NRR(I-1)
                  NRR(I-1) =  NRR(I)
                  NRR(I) = NSWAP
                  FLAG = 1
               ENDIF
            ENDDO
            FILL = FILL - 1
         ENDWHILE
         PRINT, (NRR(I), I = 1,5)
         END
```

CHAPTER FIVE PROBLEM ASSIGNMENTS

1. Write a program that will accept 10 exam scores and print out the number of A's, B's, C's, D's, and F's if

90-100	A
80-89	B
70-79	C
60-69	D
under 60	F

2. Write a program that will accept 10 measurements of adult weights and their names and sort them into *descending* order. Print the results with the names attached.

3. Modify the above program so that the average (mean) weight and the standard deviation of the weights are found.

CHAPTER 6
Structured Programming with Subroutines Function Subprograms

Structured Programming

Structured or modular programming means breaking a problem down into discrete units or modules and then programming these discrete modules to form a computer program. Each of the modules performs a specific task. For example, the last program you wrote in Chapter Five can be broken into the following discrete modules:

(1) initialize an array for names and an array for weights
(2) sort the arrays so that the weights are in descending order
(3) calculate the average weight,
and (4) calculate the standard deviation of the weights

Note: Each of the above modules can be further broken down into smaller modules.

When each of these modules are written as a ***subprogram***, the entire program is *easier to read* and *easier to modify*, and *it allows you to write new programs which include these previously written subprograms.* The two types of subprogram which we will study are the *subroutine* and the *function*.

An Introduction to Subprograms

To illustrate what a subprogram is and how it works, a somewhat crude analogy is presented. Suppose an extremely astute English student named Melville has a difficult reading assignment to complete. Now, pretend that Melville is like a computer when he begins reading. Each sentence he reads is absorbed, processed, and stored in his memory. Any time he comes across a word he does not know, he looks it up in a dictionary. So, the process Melville goes through is as follows:

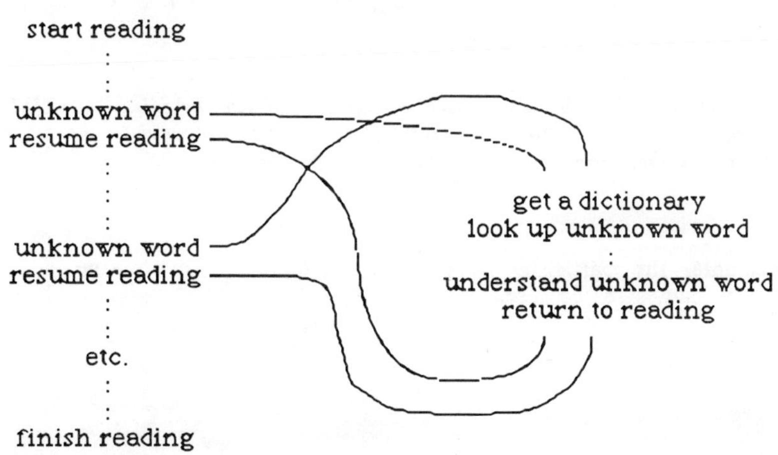

When Melville reaches an unknown word, he "calls" for the services of a dictionary. Once the meaning of the unknown word is found, he returns to his reading. This type of process is similar to a program which contains subprograms. The reading is replaced by a main program (which we have been working with since Chapter 1). The unknown word is replaced by an explicit or implicit call to a subprogram. The dictionary is replaced by a subprogram.

For example, a program which finds the average score on an exam and also uses subprograms to calculate this average looks like this:

```
* MAIN PROGRAM
*   which finds the average score on an exam
*
      REAL  SCORES(5)
      DO  I = 1,5
         READ, SCORES(I)
      ENDDO
      CALL AVERAGE (SCORES, AVG)
      PRINT, AVG
      STOP
      END

      SUBROUTINE AVERAGE (MARKS, MEAN)
      REAL  MARKS(5), MEAN
      SUM = 0.0
      DO  J = 1,5
         SUM = SUM + MARKS(J)
      ENDDO
      MEAN = SUM/5
      RETURN
      END
```

Note: *For those of you who have programmed in BASIC, the CALL statement is similar to the GOSUB statement; however, there are differences between the two (as you will see later).*

In the above example, the particular subprogram used is the subroutine. The function subprogram can also be used.

With a function, the program looks like this:

```
* MAIN PROGRAM
*   which finds the average score
*   on an exam using a function
*

        REAL SCORES(5)
        DO I = 1,5
           READ, SCORES(I)
        ENDDO
        AVG = AVERAGE (SCORES)
        PRINT, AVG
        STOP
        END

        FUNCTION AVERAGE (MARKS)
        REAL MARKS(5)
        SUM = 0.0
        DO J = 1,5
           SUM = SUM + MARKS(J)
        ENDDO
        AVERAGE = SUM/5
        RETURN
        END
```

In terms of placement, subroutines and functions in FORTRAN always come after the main program. That is,

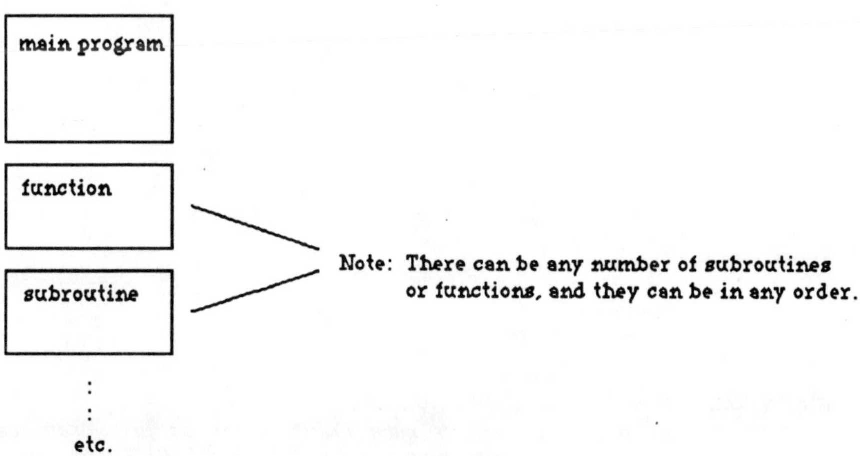

Note that each program *must* have only one main program and *may* have any number of subroutines or functions.

Subroutines

In FORTRAN, subroutines are *explicitly* called using a CALL statement. The syntax of a CALL statement is as follows:

```
CALL    <subroutine-name>  ( <argument list> )
```

where <subroutine-name> is the name of the subroutine to be called, and
<argument list> is a list of arguments (i.e., variables) separated by commas.

In the example program on page 45, the CALL statement was CALL AVERAGE (SCORES, AVG) where <subroutine-name> = AVERAGE and the argument list consists of two arguments, the array SCORES and the real variable AVG.

The general structure of a FORTRAN subroutine is as follows:

```
SUBROUTINE    <subroutine-name>  ( <parameter list> )
      :
      :
      :
RETURN
END
```

where <subroutine-name> is the name of the subroutine, and
<parameter list> is a list of parameters (i.e., variables) separated by commas.

When a particular subroutine is called, its argument list directly maps into the parameter list of the subroutine. Referring to the example program again, the array SCORES maps into the array MARKS, and the variable AVG maps into the variable MEAN.

```
          :
          :
CALL AVERAGE (SCORES, AVG)
          :
          :
STOP
END

SUBROUTINE AVERAGE (MARKS, MEAN)
          :
          :
RETURN
END
```

WARNINGS:

(1) *The number of arguments must be the same as the parameters.*

(2) *The variable types must match. That is, if SCORES is an array of reals of size 10, then MARKS must also be an array of reals of size 10. And, if AVG is a real variable, then MEAN must also be a real variable.*

Any changes to the array MARKS and the variable MEAN are passed back (or mapped back) to the SCORES array and the variable AVG, respectively.

EXAMPLE

before CALL AVERAGE (SCORES,AVG)	AVG has no value.
while in SUBROUTINE AVERAGE	MEAN = 68.4
after CALL AVERAGE (SCORES,AVG)	AVG = 68.4

Functions

In FORTRAN, functions are *implicitly* called simply by using the name of the function. The general structure of a FORTRAN function is the same as that of a FORTRAN subroutine except that the word SUBROUTINE is replaced by the word FUNCTION. When a particular function is implicitly called, its argument list directly maps into the parameter list of the function. In general, the function is used to return only *one* value, and it does this through the name of the function. Note, although the function subprogram only returns one value, it can take multiple arguments.

A function subprogram in FORTRAN is really not much different than the mathematical definition of function.

MATHEMATICAL EXAMPLE OF A FUNCTION

$z = f(a,b,c)$ where $f(a,b,c)$ is defined as $a^2 + b^2 + c^2$

FORTRAN EXAMPLES OF A FUNCTION

```
AVG = AVERAGE(SCORES)
```

where the function AVERAGE is defined as:

```
FUNCTION AVERAGE (MARKS)
REAL  MARKS(5)
SUM = 0.0
DO I = 1,5
    SUM = SUM + MARKS(I)
ENDDO
AVERAGE = SUM / 10
RETURN
END
```

Note that the value to be returned is returned thru the name of the function → AVERAGE = SUM / 10

```
Z = F(A,B,C)
```

where the function F is defined as:

```
FUNCTION F (R, S, T)
F = R**2 + S**2 + T**2
RETURN
END
```

For the mathematical example, if a=2, b=4, and c=3, then f(a,b,c) = 29. So, by replacing f(a,b,c) with 29, we get that z = 29. The FORTRAN function F operates similarly. If A=2, B=4, and C=3, then R=2, S=4, and T=3 when the function is implicitly called (by using the name of the function itself). Now, F = R**2 + S**2 + T**2 = 2**2 + 4**2 + 3**2 = 29. Hence, since F(A,B,C) = 29, then Z = 29 also.

Summary Of Subroutines And Functions

DIFFERENCES

SUBROUTINE	FUNCTION
1. explicitly called	1. implicitly called
2. returns or passes back any number of values through the parameter list	2. "returns" only one value
	3. the value returned is returned through the name of the function

Cautions

1. *When no arguments are passed to a SUBROUTINE, the parentheses for both the argument list and the parameter list are not needed because the word CALL tells the computer that the next word is the name of a subroutine. HOWEVER, when no arguments are passed to a FUNCTION, a pair of empty parentheses, that is, ( ), is needed so that the computer will recognize the word as the name of a function rather than a simple variable. If you do not understand this explanation, please ask your professor or teaching assistant for a clarification.*

2. *A subprogram can call other subprograms; but it cannot call itself either directly or indirectly.*

The Random Function

If you assign X = RND(0.) in a program, X will take on a random value between 0 and 1, exclusive (i.e., not including 0 or 1). If you assign X = RND(0.)*10, it will take on a random value between 0 and 10, exclusive, and so forth. If you assign X = INT(RDN(0.)*10), then X will take the value of a random integer between 0 and 9 (INT rounds down.).

The RND function is a little troublesome. It is not easy to come up with a random number. Random number generators often have to start with a "random" value, or they tend to produce the same number. RND(0.) in WATCOM FORTRAN takes a seed value from the last two digits of the clock in the computer. However, the AT&T 6300 with a math co-processor is so fast that, when using integers, the clock does not have time to change. Hence, it is possible to get three or four identical "random" numbers in a row. To get around this problem, we take only the initial seed from the clock and then use each successive random number as the seed for the next random number. *Be* *sure to use real zero, 0. , for the argument of RND in the original seed. Never use a seed value outside the range 0<x<1 or you will get non-random results.*

The following program will generate 10 random numbers. Type it the program and run it.

```
* RANDOM NUMBERS
*
      INTEGER  RAN(10)
      X = RND(0.)
      DO  I = 1,10
         X = RND(X)
         RAN(I)  =  INT(X*1000)
      ENDDO
      PRINT, (RAN(I), I  =  1,10)
      STOP
      END
```

CHAPTER 6 PROBLEM ASSIGNMENTS

1. Write a program which accepts 9 student names along with their exam scores. Then, in one subroutine, sort the exam scores (along with the names). In another subroutine, calculate the mean, median, and standard deviation of the exam scores. In a third subroutine, print out the sorted exam scores, the mean, the median, and the standard deviation.

2. Write a program that uses a function subprogram to calculate the values of $f(x) = x^2 + 2x + 6$ at x = 0, 1, and 10. Store these values of x and f(x) in a two-dimensional array. Then, print out the array. *Remember a function can only return one value at a time.*

3. Write a program which fills a ten element array with random integers between 0 and 100. Include a PRINT statement to check your array after filling.

CHAPTER 7
Formatting

Formatting

Until now, we have achieved output by a simple dump process using the PRINT command. However, we can exercise a considerable degree of control over the output format by using *format strings*.

The following program illustrates the use of format strings after PRINT statements. Type it in and run it.

```
CHARACTER*6   SUBJECT
N = 100
X = 98.42
SUBJECT = 'MATH'
PRINT  "( '  ' , I5)", N
PRINT  "( '  ' , F6.2)", X
PRINT  "( '  ' , E10.4)", X
PRINT  "( '  ' , A7)", SUBJECT
END
```

As shown in the example program, the format string is enclosed with double quotes (e.g., *"(' ', I5)"*). A format string contains a number of *format code(s)* separated by commas. The list of format codes is then enclosed in parentheses.

The first format code in each format string is enclosed with single quotes. It is used as a special control character for the vertical spacing of output. If the character is a blank (like all the format strings from the program), the output is single-spaced. If the character is a 0, the output is double-spaced. If the character is a -, the output is triple-spaced. And finally, if the character is a 1, the computer screen is cleared, and the output is printed at the top of the screen.

There were four format strings encountered in the example program.

(' ' , I5) indicates an integer with a width of 5 columns.

(' ' , F6.2) indicates a floating point or real number with a width of 6 columns where 2 of the 6 columns are to the right of the decimal point.

(' ' , E10.4) indicates a scientific notation number with a width of 10 columns where 4 of columns are to the right of the decimal point.

(' ' , A7) indicates a character string with a width of 7 columns.

The output for the above program should look like this:

```
ƀƀ100
ƀ98.42
0.9842E+02
ƀMATHƀƀ
```

where ƀ mean a blank space.

Notes:

1. All formatting in WATCOM FORTRAN is *right-justified* (i.e., blanks padded on the *left*). The only way to get blanks padded on the *right* of a character string is through an assignment statement or a *read* statement (see the next note).

2. Since the variable *SUBJECT* was declared to be a character string of length 6, the computer expects a character string of length 6. If the variable is assigned a string greater than length 6, then only the first 6 letters of the string are stored in the variable (the rest of the letters would be ignored). If the variable is assigned a string less than length 6, then the computer pads blanks to the right of the string so that *SUBJECT* will hold a string of length 6 (i.e., MATHƀƀ).

3. From the above notes, we see that the blank to the left of 'MATH' is from *formatting* while the 2 blanks to the right of 'MATH' are from the *assignment statement* (i.e., SUBJECT = 'MATH').

Formatting With String Constants

Type in the following program and run it.

```
*    Program using the variable type CHARACTER
*
      INTEGER    AGE
      PRINT, 'ENTER YOUR AGE'
      READ, AGE
      PRINT "( ' ' , A7, I3, A10)", 'YOU ARE ', AGE, ' YEARS OLD'
      END
```

In this program, the format string is used with the string constants, 'YOU ARE' and ' YEARS OLD'. The first format code matches up with 'YOU ARE'. The second format code matches up with the value for the variable AGE. And the third format code matches up with ' YEARS OLD'. That is,

```
PRINT "( ' ' , A7, I3, A10)", 'YOU ARE ', AGE, ' YEARS OLD'
```

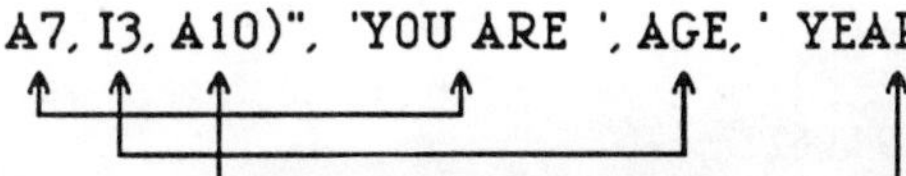

Assume AGE = 18,

YOU_ARE_18_YEARS_OLD

I	integer	E	scientific notation
F	real or floating point	X	space
A	character		

If you attempt to run any of the example programs below, make sure you include the END statement at the end of each example program.

INTEGER FORMAT:

Syntax: I*n* *or* *r*I*n*

where *n* is an integer which specifies the number of columns to allocate, and *r* is an integer which specifies the number of times to repeat a format.

Examples:

(1)
```
K = 24
PRINT "( ' ' , I5)", K
```

Output: ___24

(2)
```
I = 88
J = 94
PRINT "( ' ' , 2I3)", I, J
```

Remember that all formatting is right-justified.

Output: _88_94

(3)
```
K = 9342
PRINT "( ' ' , I3)", K
```

Asterisks are printed out if the format code does not leave enough room to print out the number

Output: ***

REAL OR FLOATING POINT FORMAT:

Syntax: F*w.d* *or* *r*F*w.d*

where *w* is an integer which specifies the number of columns to allocate, *d* is an integer which specifies the number of digits to the right of the decimal point, and *r* is an integer which specifies the number of times to repeat a format.

Examples:

(1)
```
X = 94.0622
PRINT "( ' ' , F7.3)", X
```

Output: _94.062

(2) Y = 168.42
PRINT "(' ' , F10.4)", Y

Output: __168.4200

(3) A = 24.2
B = 88.2
PRINT "(' ' , 3F5.2)", A, B, A

Output: 24.2088.2024.20

(4) X = 94.5
PRINT "(' ' , F4.2)", X

Output: ****

Asterisks are printed out if the format codes do not leave enough room to print out the number

(5) X = 94.5
PRINT "(' ', F3.1)", X

Output: ***

CHARACTER FORMAT:

Syntax: A*n* *or* *r*A*n*

where *n* is an integer which specifies the number of columns to allocate, and *r* is an integer which specifies the number of times to repeat a format.

Examples:

(1) CHARACTER*3 NAME
NAME = 'PAT'
PRINT "(' ' , A7)", NAME

Output: ____PAT

Remember that in assignment statements or READ statements, blanks are padded on the right; whereas in formatting, blanks are padded on the left.

(2) CHARACTER*4 NAME
NAME = 'PAT'
PRINT "(' ' , A7)", NAME

Output: ___PAT_

(3) CHARACTER*4 NAME
NAME = 'JENNIFER'
PRINT "(' ' , A8)", NAME

Output: ____JENN

The variable NAME was declared to be only 4 characters long. Thus, 'JENNIFER' was truncated to 'JENN'.

(4) CHARACTER*4 J, K
J = 'THE'
K = 'CLOWN'
PRINT "(' ' , 3A5)", 'BOZO', J, K

Output: _B O Z O _ T H E _ C L O W N

SCIENTIFIC NOTATION FORMAT:

Syntax: E*w*.*d* *or* *r*E*w*.*d*

where *w* is an integer which specifies the number of columns to allocate, *d* is an integer which specifies the number of digits to the right of the decimal point, and *r* is an integer which specifies the number of times to repeat a format.

Examples:

(1) X = 94.0522
PRINT "(' ' , E11.3)", X

Output: _ _ 0 . 9 4 0 E + 0 2

(2) Y = 94.0522
PRINT "(' ', E8.3)", Y

Output: * * * * * * * *

SPACE FORMAT:

Syntax: *n*X

where *n* is an integer which specifies the number of blank spaces to print out.

Examples:

(1) CHARACTER*3 NAME
NAME = 'PAT'
PRINT "(' ' , A10, 2X, A3)", 'MY NAME IS', NAME

Output: M Y _ N A M E _ I S _ _ P A T

(2) CHARACTER*5 STUDENT
STUDENT = ' DUNCE'
SCORE = 98.22
PRINT "(' ' , 2(A5, 1X), F5.1, A1)", STUDENT, 'GOT A', SCORE, '%'

Output: D U N C E _ G O T _ A _ _ 9 8 . 2 %

(3)
```
CHARACTER*5 STUDENT
STUDENT = 'DUNCE'
SCORE = 83.7
PRINT "( ' ', 2(A5, 1X), I4, A1)", STUDENT, 'GOT A', SCORE, '%'
```

Output: ERROR: Format type mismatch

Since I4 is an integer format and SCORE is a real variable, the format I4 does not match with variable SCORE. In order to correct the problem, you must use a real format (see the previous example).

__Formatting Techniques – Direct and Remote__

The following program draws your attention to the two different formatting techniques – direct and remote, the method of clearing the screen before printing, and the use of a repeated format. Remember also, that in an F format for a REAL number, such as F*w*.*d*, the *d* only includes the digits to the right of the decimal point -- it does not include the decimal point.

```
* Various Format Techniques
*
      DIMENSION RT(10), RZ(10)
      DO I = 1,10
         RT(I) = SQRT(I*10.)
         RZ(I) = RT(I) + 15.
      ENDDO

* DIRECT FORMAT with clear screen option and repeated specification.
* The '1' clears the screen. By using repeated specification, we are able
* to use '10F7.1' rather than writing 'F7.1' ten times.
*
      PRINT "('1',10F7.1)",(RT(I), I = 1,10)

* REMOTE FORMAT with repeated specification. A remote format uses
* the line number of a format expression (which may be placed
* anywhere in the program).
*
      PRINT 20, (RT(I), I = 1,10)
   20 FORMAT (10(F6.2, 2X))


* Partial listing of data with the "X" spacing technique. Each "X"
* will create one space in the output. Note the use of repeated
* specification on the contents of parentheses.
*
      PRINT 30, RT(1), RT(2), RT(3),(RT(I), I = 4,10)
   30 FORMAT (F8.3, 5X, F8.3, 5X, F8.3, 5X, 7(F8.3, 5X))


* Use of an existing format for new data (which is the main reason
* for remote format).
*
      PRINT 20, (RZ(I), I = 1,10)
      END
```

Constructing Formatted Lines Through Concatenation

Concatenation provides a useful technique for constructing formatted lines. Study the example below.

```
*    CONCATENATION EXAMPLE
     character*20 first, last
     character*40 line
     print, ' Enter your FIRST name'
     read, first
     print, ' Enter your LAST name'
     read, last
     length = lenstr(last)
     line = last(1:length)// ',   ' // first
     print, line
     end
```

Note how this example solves two problems: the variable length of the input first name and the inclusion of a comma in the output.

CHAPTER SEVEN PROBLEM ASSIGNMENTS

1. Write a program that generates ten random integers. The output should be formatted with a *remote* format, repeated specification, and spacing between values.

2. Write a program that will take as input your name and full address, and print it out nicely formatted. For example,

    Joe Schmoe
    12345 Bozo Lane
    Hollywood, CA 98989

CHAPTER 8
COMPUTER GRAPHICS - 1

In this chapter, we will learn about the possibilities and limitations WATCOM for doing graphics. In the next chapter, we will study some topics in computer graphics, such as image transformations.

Using WATCOM on an MS-DOS machine will allow you to use only the CGA mode. That is, it will display 320x200 in 4 colors (if you have a color monitor) and 640x200 in monochrome.

WATCOM Graphics

All the graphics options in WATCOM are implemented by CALLs to a library of intrinsic subroutines (the type you don't have to write yourself). Whenever we need to draw a line or a circle, for example, we do this by CALLing the appropriate subroutine. Our ability to do graphics is entirely dependent on this library unless you like creating graphics with the characters in the text mode.

Reference Materials

Most of Chapter 3 from the WATCOM FORTRAN USER'S GUIDE has been placed in Appendix A: Graphics Modes & Intrinsic Subroutines. This appendix contains a description of the color options and the graphics subroutines. A copy of the Graphics Reference Card with a summary of the graphics options is in Appendix B.

Doing Graphics

SETTING BASE VALUES.

At the beginning of a graphics program, there must be a statement setting the mode,

J=IMODE(1)	will set the medium resolution mode (med-res)
J=IMODE(2)	will set the high resolution mode (hi-res)

In hi-res, there are only two colors: black background and white foreground. In med-res, a background color must be set from a choice of 16 colors (0-15). Then, a palette (foreground color) must be chosen from a choice of two, each of which provides three colors. Thus, at any point in the program, you can choose between 4 colors (0-3), one of which is the background color. These colors are listed in both Appendices A & B. To see them, type in and run the following program.

```
*     COLOR DEMO
*     USE THE ENTER KEY FOR INPUT
         character  in
         j = imode(1)
         call  backgr(0)
         do 30 k= 0,1
            call  palett(k)
            do 20 j = 0,3
               call  boxful  (80,50,240,150,j)
               read,  in
               call  clear
20          continue
30       continue
         do  i = 0,7
            call  backgr(i)
            read,  in
            call  backgr(i+8)
            read,  in
         enddo
         stop
         end
```

Proceed through this program using the ENTER key. The program begins by paging through the colors of the two palettes. (In the first palette [palett(0)], the last color is yellow. It is, however, referred to as brown in the documentation.) Press the ENTER key to advance the program through the colors. After the palettes, you will go through the background colors which are full-screen. The background colors come in 8 basic colors (including black and white) and 8 light colors (including gray and bright white). The program has paired the basic and light colors together and, as you will see, the difference is small.

The last time you press ENTER should bring you back to the program screen. Examine the program. The MODE is set to 1 and the background to 0 (black). The DO 30 loop invokes the two palettes and the DO 20 loop uses the BOXFUL subroutine to represent the colors. BOXFUL produces a rectangle described by the coordinates of opposite corners and filled with a color specified by the last parameter. The subroutine CLEAR clears the screen to the background color. Note the use of the READ statement; it will accept the ENTER key by itself. You do not have to input a character and then use ENTER.

To use the subroutines, you will need to learn what the arguments represent. To do this, see Appendix A.

Images

There are many ways to draw shapes directly. BOX draws empty rectangles, BOXFUL draws filled rectangles, LINE draws lines, CIRCLE draws circles, and so on. In fact, it is quite possible to write a graphics program that is simply a list of CALLS. Use *D to clear your program space and enter the following program.

```
*   ARC AND PAINT DEMO
        integer    j
        j = imode(1)
        call  backgr(1)
        call  palett(0)
        call  arc(160,100,60,20,1)
        call  arc(200,50,30,10,1,0,90)
        call  arc(200,50,30,10,2,90,360)
        call  paint(160,100,3)
        end
```

This program demonstrates the ARC and PAINT subroutines. The ARC subroutine will draw an ellipse, or any part of it. When the X and Y diameters are the same, you will get a circle or part of it. In the program, the first use of ARC generates a full ellipse. In the next two uses of ARC, a new ellipse is generated of two different colored arcs. The PAINT command starts with the point specified by the first two parameters and fills the nearest border with the color specified by the last parameter.

Parameters

The arguments of the subroutines typically take integers (i.e., actual integer numbers or integer variables). When plotting on the screen in raster graphics, only integers are used since the screen image is composed of pixels (integer coordinate pairs). No fractional positions can be represented. If you prefer to work with X and Y as variables, be sure to declare them as INTEGERs. Clear your program space, and type in the program below.

```
*   DEMO FOR BOX, BOXFUL, AND PUTDOT
        integer  q
        j = imode(1)
        call  clear
        call  backgr(1)
        call  palett(0)
        do 20 i = 10,160,15
            call  box  (2*i,i,2*i+20,i+10,2)
            call  boxful  (i,i,i+10,i+30,1)
            q = i **1.2
            call  putdot(q,q-i,1)
20      continue
        stop
        end
```

This program shows an example for BOX, BOXFUL and PUTDOT. The PUTDOT subroutine simply plots a single point.

Line Drawing

One of the most general images is the straight line. The LINE subroutine does this. Doing drawings line by line is a slow process, but there are some ways to speed this up. The DRAW command will draw a character string, where in a shape is described, in the form of "R50u100", etc., where "R50", for example, means draw a line to the right, 50 long, from the current point. Not only is each line easier to draw this way, but the shape as a whole may be saved as a string and specified by name only to be reproduced elsewhere. One of the reasons this is more efficient is that it is using relative coordinates for specification. The only absolute specification is the starting point for the string. Another technique to speed up the generation of lines is by putting the LINE subroutine in a loop and using variables as arguments. This is only possible where the lines have a systematic relationship to each other. To see examples of this discussion, type in the following program and run it.

```
*   PROGRAM LINE DEMONSTRATION
        integer  y1, x2
        character  pic*30
        j = imode(1)
        call  backgr(0)
        call  clear
        call  palett(1)
        pic="bm100,50r120e100L120u100c3"
        call  draw(pic)
        do i = 1,10
            y1 = 45 + 10 * i
            x2 = 95 + 12 * i
            call  line(100,y1,x2,150,5)
            call  line(220,y1,x2,50,6)
        enddo
        stop
        end
```

Reproduction

The easy replication of images is one of the most powerful features of computer graphics. There are a number of ways to achieve it, some more subtle than others. One of the least subtle methods is using the GETPIC and PUTPIC subroutines. In this technique, you simply save everything being displayed, including the background, in a rectangle of the screen (i.e., GETPIC); then you can bring it back, and put it wherever you want it (i.e., PUTPIC). In order to do this, a very large character string is needed to store the picture in. Type in and run the following program.

```
*    DEMO FOR DRAW AND PUTPIC
        integer  y1, x2
        character  pic*30,  sq*4000
        j = imode(1)
        call  backgr(0)
        call  clear
        call  palett(1)
        pic="bm100,50r120d100L120u100c3"
        call  draw(pic)
        do i = 1,10
           y1 = 45 + 10*i
           x2 = 95 + 12 * i
           call  line(100,y1,x2,150,5)
           call  line(220,y1,x2,50,6)
        enddo
        call  getpic(sq,100,50,220,150)
        do i = 10,90,20
           call  putpic(i,i,sq,0)
        enddo
        stop
        end
```

If you examine this program, you will see that it is an extension of the Line Demonstration program. We save the picture with GETPIC by specifying the rectangle of the screen which contained what we had drawn. This picture is then stored in a character string called SQ which has 4000 characters assigned to it in the declaration statement. The rectangle is 120x100, that is 12000 pixels. These 12000 bits of information are stored 4 bits/byte so we declare SQ*4000 to be on the safe side. In hi-res, we would have to double the storage because the resolution is twice as high. You will notice that GETPIC stores the image and PUTPIC retrieves it. This is the reverse of the meaning of the PUT and GET commands for file management.

The PUTPIC does have a relatively artful feature. The last parameter takes the values 0-4. The five possibilities involve a pixel-by-pixel comparison of the image to be imposed and the existing image (excluding background): (1) replace, (2) replace with the negative of the new image, (3) invert non-zero points (good for animation because you can draw and then erase your image), (4) show only points common to both, or (5) superimpose leaving in place any parts of the existing image not covered.

<u>High Resolution</u>

All the subroutines will work in hi-res also, but the color specification is reduced to black(0) and white(1). The demonstration program also includes an example of the use of the subroutine for drawing circles.

```
*    CIRCLE DEMONSTRATION
        integer x, y, r
        j = imode(2)
        call clear
        x = 80
        y = 50
        r = 20
        do 20 i = 1, 10
           call circle (x,y,r)
           x = x + 18
           y = y + 7
           r = r + 10
20      continue
        stop
        end
```

<u>COMPUTER GRAPHICS - 1 PROBLEM ASSIGNMENTS</u>

Write a program that will move a circle along a sine curve. Provide an option for leaving the circles in place or for erasing as you go. For an extra thrill, try modifying the radius as you go.

CHAPTER 9
COMPUTER GRAPHICS - 2

This chapter consists of a short discussion of some fundamental concepts in computer graphics followed by an introduction to image transformation.

Raster Graphics

Almost all computer graphics are generated by sweeping an electron beam over a phosphor-backed screen. The beam is turned on where an image is desired. In vector graphics, the line trace is continuous but only the image is traced - nothing else on the screen is touched. In raster graphics, the beam sweeps the whole screen line by line and each horizontal line is treated as many discrete points or pixels (picture elements). Each pixel is lit or unlit according to the desired image. The final image in raster graphics is composed of many points, thus a diagonal line will appear jagged. The more pixels per inch you have, the better the image. A resolution of 320x200 is not bad, but 640x400 is four times as good.

Vector graphics has smooth lines, and line for line appears far superior to most raster images. However, for both vector and raster graphics, the image needs to be refreshed about 50 times a second to avoid flicker and image decay. For complex shapes, particularly with solid areas filled in, vector graphics is very hard to achieve. Thus, in this case, raster graphics is preferred. In addition, in terms of cost and versatility, raster graphics has proven to be cheaper and more versatile.

If the resolution of a raster display is 320x200, there are 64,000 pixels to be controlled. This is often done by bit-mapping where each pixel has a corresponding bit in memory which contains the information "on" or "off". A bit is the fundamental memory component in a computer. It may take two values only, 1 or 0; or in electrical terms, high or low voltage. This is why computers work with a binary number system. There are 8 bits in a byte so the 64,000 pixels in our example would need 8,000 bytes or 8K of memory to store the image. A 640x400 resolution needs 32K to store the image. Rapidly falling memory costs have played a big role in increasing the resolution and popularity of raster graphics.

Coordinates

The size and shape of an object is usually described in terms of its dimensions (e.g., 10 feet long by 4 feet wide). In computer graphics, we plot the shape, which means describing it in terms of its coordinates. So, a shape which is 10 feet long and 4 feet wide can be described as (0,0); (10,0); (10,4); and (0,4). This places a corner of the rectangle at the origin. We could place the shape elsewhere by adding a constant to each X coordinate, or to each Y coordinate, or to both. The shape is drawn by joining the points. Thus, drawing a shape involves two steps: (1) defining the shape in coordinate terms, and (2) defining the relationship between the shape and the origin of the coordinate system.

Screen Coordinates

In general, the global system of coordinates which define a shape will be different from the computer screen coordinates. The global system of coordinates is what you usually work with when you are drawing graphs by hand; that is, the origin is usually at the bottom left corner. In contrast, screen coordinates typically have the origin at the upper left corner of the computer screen. Positive X-coordinates move from the left hand side of the screen to the right hand side, and positive Y-coordinates move from the top of the screen to the bottom of the screen.

Because of the differences between global coordinates and screen coordinates, you will have to convert from global coordinates to screen coordinates in order to display your image. This may often be quite easy; but, in particular, you must remember that Y is measured downwards in screen coordinates. Thus, you may need a line that converts the global Y-coordinates to screen Y-coordinates -- such as Y(I)=199-Y(I). This would, in effect, place the origin at the bottom left of the screen rather than the top left.

Another obvious conversion will be reconsidering the relationship of the shape to the origin; that is, the screen and global origin may be different. This adjustment, if required, means adding or subtracting a constant from all the X-coordinates and doing the same, although probably with a different constant, for the Y-coordinates.

In converting to screen coordinates, you need to consider the scale of conversion. Your image may be too large or too small when compared to the size of the screen which in med-res, for example is 0 to 319 (X-axis) and 0 to 199 (Y-axis). This conversion is done by using the ratios of GLOBAL RANGE OF X/SCREEN RANGE OF X and GLOBAL RANGE OF Y/SCREEN RANGE OF Y.

Since raster graphics only uses integers, your conversion to screen coordinates must involve the conversion of any real numbers to integers.

Absolute and Relative Plotting

In the last chapter, we saw how the DRAW subroutine involves moves made relative to the last position. This relative coordinate specification can also occur when you are making repeated transformations of a shape, as you will discover later in this chapter. That is, you must decide whether you wish to transform the current shape from its current location or the original shape from its original location. Working with the original shape will mean storing the original descriptive information and using a new set of variables to generate the tranformed shapes and locations.

Animation

We will not be working with animation techniques explicitly, but a couple of tips may be worthwhile. Animation refers to the simulated motion of a shape on the screen. It is achieved by drawing and erasing a shape along a "motion" path. There are several techniques for doing this. Plotting in a foreground color will display the image; plotting the same image in the background color will erase the image. The PUTPIC and GETPIC subroutines offer another means of doing this. Using CLEAR to clear the whole screen is a little drastic, but it may, on occasion, be the best choice for erasure. It is important to structure your program so that display time is as long as possible relative to the erased time. Preferably, display should occur immediately after erasure.

The distance "travelled" between successive displays on the screen is one of the determinants of the "speed" at which the shape appears to move. In speeding up, your shape you will reach a point where a "kangaroo" effect of jumping rather than gliding will occur. Unless you image is a kangaroo or a flea, this jumping effect is to be avoided.

Image Transformation

One of the most powerful and pleasurable features of computer graphics is the ability to transform images in an infinite number of ways. We will study replication, translation, scaling, and rotation of images. All of these, and several others as well, may be expressed in two equations.

$$X'=A*X+B*Y+C \qquad [1]$$
$$Y'=D*X+E*Y+F \qquad [2]$$

In these equations, X and Y are the existing coordinates of a point, and X' and Y' are the coordinates of the point after transformation. Different values of the constants A,B,C,D,E and F will produce different transformations. When each point in a shape is subjected to the same transformation, the whole shape will be transformed.

Image Translation

This is the simplest transformation. It simply means moving the image. In terms of our two equations, we have A=E=1 and B=D=0.

$$X'=X+C \qquad [3]$$
$$Y'=Y+F \qquad [4]$$

The translation effect depends on the values of C and F. For example, if C=20 and F=0, the shape will be moved 20 to the right when this transformation is applied to each point in the shape. If you only applied the transformation to one point on a shapes, you would achieve an effect known as dragging.

```
*   DEMONSTRATION OF X AND Y TRANSLATION
        integer x1, y1, x2, y2,  inx, iny
        character  ps*1
        j=imode(2)
        ps = 'p'
        while (ps .ne. 'q')
           print, "Enter two integers, the first < 600,  the second < 150"
           read, inx, iny
           x1 = inx + 50
           y1 = iny + 20
           x2 = inx + 100
           y2 = iny + 50
           call  box(x1,y1,x2,y2)
           print, 'Press any key to continue, q to quit'
           read, ps
        endwhile
        end
```

When analyzing graphics programs, it is often easiest to look first for the "output" statements. In this program, the BOX subroutine creates output to the screen, so the program draws rectangles. Now, examine the arguments of BOX -- that is, x1, y1, x2, and y2. If you look at the assignment statements that control these arguments, you will see equations [3] and [4] as clearly as you will ever see them. The rectangle is being translated, with INCX and INY being equivalent to C and F in equations [3] and [4] respectively. The BOX drawing subroutine is structured in a WHILE loop which will keep accepting new translation coordinates until you enter a "Q" for the second READ statement. Pressing any other key will keep the loop going. Note that this program always modifies the original shape. Now, take a moment to figure out how the program may be changed so that translation is done relative to the last position drawn.

Replication

This is really a variation on translation, scaling, or any other transformation. It means drawing a shape repeatedly. It differs from animation because there is no erasure. The purpose may be to achieve a desired pattern, or maybe to create a pseudo-3D effect. In the last chapter, we provided a program for drawing circles in hi-res. If you look at this program, you should realize that by repeatedly drawing a circle the created image appears to have three dimensions. This 3D effect is enhanced by a gradual enlargening of the image to give it a perspective effect. The perspective effect can be improved by adjusting the spacing between each drawing.

Scaling

If we have an object 10 long and 5 feet wide, and then double its size, it will be 20 feet long and 10 feet wide. If the object is described in coordinate terms we have a problem. Consider the same object placed at (10,10); (20,10); (20,15); and (10,15). If we double all these coordinates, we will translate the object as well as double it in size. The result will be (20,20); (40,20); (40,30); and (20,30). To see why, we can look at equations [1] and [2] for the case of scaling (B = C = D = F = 0). Thus, equations [1] and [2] simplify into the following equations.

$$X' = A*X \qquad [5]$$
$$Y' = E*Y \qquad [6]$$

Equations [5] and [6] show that all X coordinates are increased or decreased by the factor A, all Y coordinates by the factor E. The only coordinates not affected by scaling are those of the origin, (0,0). The origin is the one point that will stay in the same place after scaling. Therefore, you must decide which point you wish to remain invariant and place the origin there, at least for the duration of the scaling transformation. For our example, suppose we wish to make the bottom left corner, (10,10), invariant. To do this, we must subtract (10,10) from all the coordinates so that they become, (0,0); (10,0); (10,5); and (0,5). Clearly, the bottom left corner is now the origin and will not be affected by the transformation. Next, we scale by doubling the coordinates so that they become (0,0); (20,0); (20,10); and (0,10). Finally, we must add back the (10,10) to put the origin back where it was. Thus, we end up with the correct coordinates of (10,10); (30,10); (30,20); and (10,20). The algebraic sequence is as follows,

$$X' = X - OX$$
$$Y' = Y - OY$$
$$X'' = A*X'$$
$$Y'' = E * Y'$$
$$X''' = X'' + OX$$
$$Y''' = Y'' + OY$$

Or alternatively,

$$X' = A*(X - OX) + OX \qquad [7]$$
$$Y' = E *(Y - OY) + OY \qquad [8]$$

The invariant point is represented in these equations by OX and OY.

To see an example of a scaling program that does not include this adjustment and hence suffers from the translation effect, run the following program.

```
*   SCALE DEMONSTRATION
      integer x1, y1, x2, y2, in
      character  ps*1
      j = imode(2)
      ps = 'p'
      while (ps .ne. 'q')
         print, 'Enter an integer less than 10'
         read, in
         x1 = in * 50
         y1 = in * 20
         x2 = in * 100
         y2 = in * 50
         call box (x1,y1,x2,y2)
         read, ps
      endwhile
      end
```

Notice that the program bears a strong resemblance to the translation program.

The SCALE program runs on integers, but it does not have to. It is possible to use scaling factors such as 0.1 or 13.85. If you write a program to do this, just remember to convert to integers before plotting.

In the SCALE DEMONSTRATION, the scale factors for X and Y were the same (i.e., the value of the variable *in*). Alternatively, another possibility is to use different scale factors for X and Y. The last program has been modified to do this,

```
*   SCALE X AND Y DEMONSTRATION
        integer x1, y1, x2, y2, inx, iny
        character  ps*1
        j = imode(2)
        ps = 'p'
        while (ps .ne. 'q')
           print, 'Enter two integers less than 10'
           read, inx, iny
           x1 = inx * 50
           y1 = iny * 20
           x2 = inx * 100
           y2 = iny * 50
           call box (x1,y1,x2,y2)
           read, ps
        endwhile
        end
```

Like the previous version, this program has no invariant point feature. The advantage of using different scale factors for X and Y is that is allows you to change the shape of the image.

After you have run the SCALE X AND Y DEMONSTRATION program several times and gotten a sense of how radically the shape may be changed, modify the program to include an invariant point feature.

Rotation

This is the most complex transformation that we will examine. Even so the same two equations (equations [1] and [2]) prevail, becoming in this case,

$$TEMP = COS(AN)*X - SIN(AN)*Y \quad [9]$$
$$Y' = SIN(AN)*X + COS(AN)*Y \quad [10]$$
$$X' = TEMP \quad [11]$$

In these equations, AN is the angle of rotation, clockwise positive. If you know your trigonometry, they are not hard to derive, and you might like to do it on scrap paper.

This is the first time that we have considered a transformation that has non-zero values for B and D of equations [1] and [2]. The consequence is important. Since the new values X' and Y' are a function of both the old X and the old Y, we must be careful not to generate the new X and then calculate the new Y from the old Y and the new X. If we give the new coordinates new variable names, there will be no problem. But very often we will be using the same variable name, and we will simply be writing X for the X' above. In this case, a temporary variable is necessary to hold the new value of X until the new value of Y has also been calculated.

The translation problem of the scaling transformation is also a problem for rotation. The only point not affected by rotation is the origin, as before. All other points are inclined to fly off at an angle. Once again, you must choose an invariant point, an origin for the duration of the transformation. In this case, it is the point about which everything else will rotate. Thus, the equations need to be modified as follows.

```
TEMP = COS(AN)*(X-OX) - SIN(AN)*(Y - OY)     [12]
Y' = SIN(AN)*(X-OX) + COS(AN)*(Y - OY)       [13]
X' = TEMP                                    [14]
```

To see an example of a program which will rotate a shape input in coordinate pairs, type in the following program.

```
*    ROTATION DEMO
         character  ps*1
         integer  n
         real  ex(20),  wy(20)
         j  = imode(2)
         ps = "p"
         while  (ps.ne."q")
            call  shape  (ex,wy,n)
            call   rotation(ex,wy,an,n)
            call  clear
            call  plot(ex,wy,n)
            read,  ps
         endwhile
         end

*    DATA INPUT
         subroutine   shape(ex,wy,n)
         real  ex(20),  wy(20)
         integer  n
         print,  "how  many  points?"
         read,  n
         print,  "enter  the  coordinate  pairs  -x,y  enter  -  for  each  point"
         read,   (ex(i),wy(i),i=1,n)
         return
         end

*    ROTATION
         subroutine    rotation(ex,wy,an,n)
         real   ex(20),wy(20)
```

(continued on next page)

```
      integer  n
      real  x,y,  an
      print,  "Enter  coordinates  of  the  point  of  rotation"
      read,  x,  y
      print,  "Enter  the  angle  of  rotation"
      read,  an
      an = an * 3.14159 / 180.
      do i = 1,n
         ex(i) = ex(i) - x
         wx(i) = wx(i) - y
         temp  =  ex(i)*cos(an)  -  wy(i)*sin(an)
         wy(i)  =  ex(i)*sin(an)  +  wy(i)*cos(an)
         ex(i) = temp + x
         wy(i) = wy(i) + y
      enddo
      return
      end

*  PLOTTING
      subroutine  plot(ex,wy,n)
      real  ex(20),wy(20)
      integer  n
      call  putdot(int(ex(1)),int(wy(1)))
      do i = 2,n
         call  line(int(ex(i-1)),int(wy(i-1)),int(ex(i)),int(wy(i)))
      return
      end
```

The WHILE loop in the main program mainly consists of calls to subroutines to input the shape, rotate the shape, and plot the shape. To run the program the first time, try entering a straight line, say, (100,100) and (200,100). Rotate it 30 degrees around the point (100,100). After this, you can experiment freely with it. Examine the separate effects of variations in the angle of rotation and the point of rotation. One thing you will quickly realize is that it would be nice if once a shape were loaded it would stay there. The problem assignments for this chpater were written assuming you would want to do something about this.

COMPUTER GRAPHICS - 2 PROBLEM ASSIGNMENTS

1. Modify the ROTATION program so that you can continue rotating the input shape form the last position it was in. This requires only a few simple changes. The key is the WHILE loop of the main program.

2. Modify the ROTATION program so that the user can keep trying different rotations on the original shape in their original location. This will mean adding new arrays to store the original data in.

CHAPTER 10
THE DISPLAY AND ANALYSIS OF MATH FUNCTIONS

For as far back as we know, the history of mathematics has shown an intimate interplay between spatial and numerical formulations. For a while, the Pythagorean school held that "all was number", but by about 400 BC this position had fallen to challenges from geometry -- not without some trauma for the Pythagoreans. Even earlier, the "rope-stretching" Egyptians and the Babylonians had mastered elements of geometry and surveying. (The Egyptians used surveying to reestablish property lines on the flood plains of the Nile after each annual flooding.) Both of them also had established a system of numbers adequate for rudimentary arithmetic, and both had developed primitive forms of algebra and geometry which were symbiotic. That is, the one was stimulated by and, in turn, affected by the other.

The relationship goes back much further -- well before recorded history, in fact. Evidence of counting has been found dating from the Neanderthal period 50,000 years ago, and the following group of counting designs dates from Paleolithic time, 25,000 years ago. If you look at it, it is hard to ignore the relatedness of the three designs (Egyptian/Babylonian, Neanderthal, and Paleolithic). If you wrote down a corresponding numerical series for the group of pictures below, what would it be? It is difficult to see this group as purely an exercise in drawing since it has such a deliberately abstract mathematic quality to it.

From 400 BC until 529 AD, when Justinian, the Eastern Emperor, closed the schools at Athens to "protect" Christianity from their pagan influences, the Greeks had a continuous tradition in mathematics. At times, this was spread around the Mediterranean and it certainly involved interactions with Egypt, Mesopotamia and probably India and China. The developments were enormous. Geometry and algebra developed into symbolic systems of knowledge that were not restricted to utility, as were the earlier Egyptian and Babylonian efforts to a large degree. From this, came planar and solid geometry, conics, trigonometry and algebra. Much of the algebra developed from geometric formulations as some of the examples for which we will write programs indicate.

Seeing is Knowing

When all you know of a math function is its algebraic form, it may be hard to remember. To be honest, when you look at some of them, they may be hard to like, also. This is because many math functions are a little tricky to analyze and fairly cumbersome to work with.

Graphing a function solves both these problems. The graph of a function is usually much easier to remember than its algebraic form. This is because so much numerical information is represented by a pattern. Further, in a language-dominated culture, it is good to give the visual capacity of the right brain something to do and to relieve the overburdened left brain.

Although it may take some manipulations of scale, the graph of a function will display its roots, its points of zero slope (in calculus this provides the roots of the function in its first derivative form), its points of reversing curvature (in calculus this provides the roots of the function in its second derivative form), its values at X=0 and infinity and its asymptotes. All the values derived visually from the graph will only be correct to a limited number of significant figures, and this tends to be particularly true of the points of zero slope and inflection (reversing curvature). However, if you don't mind continually zooming your graph, you can, in principle, get a root to as many significant figures as you want. For example, you could repeatedly zoom your graph until you could see a root at approximately the midpoint between, say, 3.4765 and 3.4766. This would give you an estimate for the root of 3.47655.

Consider the following function. On the AB calculus advanced placement exam in May of 1988, students were asked to sketch it. Using calculus, this function may be sketched. However, without any knowledge of calculus, the function may be plotted by substituting a series of values for X and calculating the results. This could be done by hand. As it happens, this is precisely the kind of drudgery that led mathematicians, scientists and engineers throughout history to invent calculating machines, from the abacus to the electronic computer.

```
* GRAPH OF THE FUNCTION Y=X*EXP(-X)
* BY INSPECTION, Y=0 WHEN X=0 AND Y->-∞ WHEN X-> -∞.
*
* BY GRAPHING OR CALCULUS, THE SLOPE = 0 AT X=1, Y=0 IS AN ASYMPTOTE,
* AND THE CURVATURE REVERSES AT X=2
*
          integer x,y
          real i
          j = imode(2)
* Draw axes to be centered in a screen of size 640(x) by 200(y)
          call line (320,0,320,200)
          call line (0,100,640,100)
* Calculate and plot the function for -3 < x < 3. Note that the origin is placed
* at (320,100). Since the screen plots y positive downward, the y value is
* calculated with a reverse sign to cancel this effect.
          do i = -3,3,.2
            if (i.eq.0) then i = 0.001
            x = int(i*100) + 320
            z = 20*i*exp(-i)
            y = -int(z*1) + 100
            call putdot(x,y)
          enddo
* Make hash mark at x=1, y=0 and x=0, y = 1
          call line (420,970,420,103)
          call line (315,70,325,70)
          end
```

The following function is a little unusual, and graphing it is very helpful. When analyzed by standard calculus, you find one point of zero slope, an indefinite second derivative, and two points where the function equals zero. This is a bit puzzling, and all the found values are very small. As it happens, on a large scale, the graph is roughly a straight line. However, on a small scale, plotting the range -0.03 < x < +0.03 reveals the uniqueness of the function. It turns out to be anything but a straight line in this range.

After running the program as listed, modify it to display the function for the range -10 < x < +10 to see how easy it is to miss the critical aspects of a function. Doing this involves changing the "do" statement and the assignment statements for x and y.

```
* Graph of the function y - x**(2/3) - 4*x
*
        integer x,y
        j = imode(2)
        call line (0,100,640,100)
        call line (320,0,320,200)
        do t = - 0.03, 0.03, 0.0002
           p = t**2
           q = p ** (1./3.) - 4.0 * t
           x = int (t*10000.) + 320
           y = -int(q*1000.) + 100
           call putdot(x,y)
        enddo
        end
```

********** Problems? **********

Note that the initial statement in the DO/ENDDO loop may be written as,

```
do t = a, b, c
```

and a, b and c could be input from the keyboard. This would allow you to zoom the graph, at least for X over a specified range. In the above program we have coded in the zoom that we want. Rewrite the program to allow the user to control the zoom. You will find that three problems emerge that require resolution.

1. The program needs to let the user to keep trying different input values. This may be done in a WHILE/ENDWHILE loop with the conditional value being input at the end of the loop each time (e.g., "Do you wish to continue? Y/N").

2. You need to decide about the axes. Will you recalculate where they should be plotted (if they appear at all), or will you leave them as they are and reinterpret them for each range of X plotted? If you choose the latter, which is our preference, where do you want the axes? Remember, the axes are just reference lines. What about divisional hash marks?

3. As you zoom on an increasingly smaller X range, the changes in Y also get smaller. It may be necessary to arrange for a modification of the Y scale, also.

Parametric Programming

Many awkward math functions in x and y can be expressed by two simple expressions of both x and y as a function of a new parameter. Sometimes these parametric expressions are a consequence of the way the expression is derived. This is particularly true of those developed by the Greeks, where the two equations often represent the locus of a point under specified constraints. Sometimes the parameters are introduced because they simplify. However, there are no set rules or methods for discovering them and it is not always obvious. In the following example, a parameter, s, is used in conjunction with two constants to establish a pair of expressions which represent an ellipse. The constants represent the major and minor diameters of an ellipse, so you can get any ellipse by varying these constants (including the limiting cases of a circle and a straight line).

```
*  Parametric programming and plotting of an ellipse.
*  The constants a, and b, are taken to be greater than one for practical,
*  not theoretical, reasons.  The computer cannot display quantities
*  less than an integer.
*
           integer a, b, x, y
           j = imode(2)
           print, 'please enter integer coefficients A, B greater than 1'
           read, a, b
*  Fudge factor for the screen ratio
           a = int(2.35*a)
*  The index, s, is incremented through a 2 pi range and the function is
*  plotted.
           do s = 0, 6.3, .1
              x = int(a*cos(s)) + 320
              y = int(b*sin(s)) + 100
              call  putdot(x,y)
           enddo
           end
```

********* Problems? **********

Now, we can draw ellipses and circles. But why settle for only one at a time? Modify the program so that it will draw 10 ellipses each time. Do this in two different ways. First, do it by producing a family of like ellipses. That is, by keeping the a/b ratio the same but changing the size. (You could also keep the same ellipse and change its position on the screen, e.g., place a series of ellipse in a circular pattern.) Next, do it by keeping one diameter the same and varying the other diameter.

Now, produce a program that lets the user choose whether to produce a family of like ellipses, or a family of variable ellipses with one diameter constant. There is a restriction: you can only use one coding of the calculating and plotting routine. This means putting it in a subroutine and structuring the alternative solutions in the main program.

There is a fundamentally playful quality about computer graphics. You are welcome to move on at this point, but some of you will want to try more complex variations first, such as, plotting a series of ellipses along a sine curve with the ellipses gradually changing in a/b ratio and increasing in size.

More Parametric Programming

One of the ancient Greeks who enjoy geometry was Diocles who thought this cissoid interesting because it solved a problem in proportions. It is one of those math functions which was first expressed as two separate parametric equations which represented the path of a point.

```
*           CISSOID OF DIOCLES:  (Y**2)(2*A - X) = x**3
*           SUBSTITUTE:  Y= T*X, GIVING X = 2*A*T**2/(1+T**2)
*
            integer  x,y
            j = imode(2)
            a = 150.
            do t = -1,1, 0.1
               q  =  2*a*t**2/(1+t**2)
               x = int(q) + 320
               r = t*(x-320)
               y = int(r) + 100
               call  putdot(x,y)
            enddo
            end
```

And now, in glorious low-res color,

```
*           CISSOID OF DIOCLES:       (Y**2)(2*A-X) = X**3
*           SUBSTITUTE:   Y = T*X,  GIVING X = 2*A*T**2/(1+T**2)
*           LOW RESOLUTION COLOR MODE
            integer  x,y
            j = imode (1)
            call  backgr(3)
            call  foregr(2)
            a = 150.
            do t = -1,1, 0.05
               q = 2 * a * t**2 / (1 + t**2)
               x = int(q) + 160
               r = t * (x-160)
               y = int(r) + 100
               call  putdot(x,y)
            enddo
            end
```

The hypocycloid below is produced by the path of a point on a circle rolling around the inside of a larger circle. It, too, was derived in two parametric equations from the geometric constraints.

```
*        HYPOCYCLOID OF FOUR CUSPS:  X**(2/3)+Y**(2/3) = A**(2/3)
*        SUBSTITUTE     X = A * (SIN(T)**3), Giving
*                        Y = A * (1-SIN(T)**2)**(3/2)
         integer  x,y
         j = imode(2)
         a = 75.
         do t = 0,1.6,0.1
            p = sin(t)
            q = a*p**3
            x = int(q) + 320
            r = a * (1. - p**2) **(3/2)
            y = int(r) + 100
            call  putdot(x.y)
            y = -int(r) + 100
            call  putdot(x,y)
            x = -int(q) + 320
            call  putdot(x,y)
            y = int(r) + 100
            call  putdot(x,y)
         enddo
         end
```

********* Problems? **********

1. Draw four of these hypocycloids with two cusps from each touching two other cusps.

2. The following expression will graph as a "rose". (Actually, it is a four petal pattern more like the flower of a phlox plant.) You are given the expression in polar coordinates. To plot it you will need to reduce it to Cartesian coordinates.. When you have plotted a rose successfully, produce a series of them on circular path.

   Four-leaved rose:

   s = 0.5*(a-b)*sin(2*t)

   where s is the distance from the origin to the plotted point, t is the angle a line from the origin to the plotted point makes with the positive X axis and a and b are constants.

********* ADDITIONAL PROBLEMS **********

1. The cardioid: s = a*(1+cos(t))

2. The involute of a circle (formed by the end of a piece of string when it is unwound from a fixed circle of radius r):

   x = r*cos(t) + r*t*sin(t)
   y = r*sin(t) - r*t*cos(t)

3. The folium of Descartes:

```
x**3 + y**3 - 3*a*x*y = 0
```

substitute y = t*x and obtain x as a function of (a,t) and y as a function of (a,t).

4. Instead of plotting points with "putdot" use "line(x1,y1,x2,y2)" to draw one of the functions with lines. (Hint: you will have to save the last points calculated for one loop each time.)

5. Choose your own nasty function and graph it.

APPENDIX A

GRAPHICS MODES

&

INTRINSIC SUBROUTINES

Chapter 3

Intrinsic Functions and Subroutines

3.1 IBM Personal Computer Display Support

A library of intrinsic functions and subroutines supports a variety of display options for both the Color/Graphics Monitor and the Monochrome Display. As well, this library includes features designed to support drawing of graphic images with colour. These features are described together in this single section for convenience of reference.

3.1.1 Introduction

Various display colours, attributes and modes can be selected for the IBM Personal Computer. The Color/Graphics Monitor supports a broad selection, whereas there are fewer choices for the Monochrome Display. In all cases, there is a *background* colour and a *foreground* colour. The *background* is the colour of the screen where it is blank; that is, where no text characters or graphic images are displayed. The *foreground* is the colour of *text characters* displayed. Text characters, such as letters and digits, are normally displayed with *WRITE* (or *PRINT*) statements from a FORTRAN program, or in response to typing on the keyboard.

There are three modes of display supported for the Color/Graphics Monitor: *text*, *medium-resolution graphics* and *high-resolution graphics*. The two graphics modes are referenced herein as *medium* and *high* for brevity. The particulars of the various modes are explained in subsequent subsections of this section. In text mode with the Color/Graphics Monitor, nothing is displayed in the outer perimeter area of the screen known as the *border*. A unique colour can be chosen for this border. In the two graphics modes, a colour can be selected for *graphic images*. Colour choices for background, foreground, border and graphic images are described in the section entitled "Colour Choices".

Only text mode is supported for the Monochrome Display. Combinations of foreground and background "colours" support display of characters in normal image (white on black) and reverse image (black on white) and as blinking, underlined, bright (high intensity) or invisibile. These capabilities are described in the section entitled "Monochrome Display Attributes".

Screen *width* can be specified as 80 or 40 character positions in text mode. In graphics modes, text characters are displayed with width 40 using

medium resolution and width 80 using high resolution. The screen width is always 80 for the Monochrome Display and cannot be altered. In all modes, text characters are displayed in 25 lines on the screen.

3.1.2 Text Mode

In text mode, text characters can be displayed with individual attributes for each character on the screen. That is, each character position on the screen can have individually specified foreground and background settings. The attributes of a character correspond to the foreground and background settings when the character was displayed. Changes to the foreground and background settings effect subsequent characters displayed, but do not change the attributes of characters already on the screen.

Character attribute options are described in the sections entitled "Monochrome Display Attributes" and "Colour Choices". These sections apply to the use of the Monochrome Display and the Color/Graphics Monitor, respectively.

When using the Color/Graphics Monitor, the widest choice of colours is available in text mode. Certain characters in the Screen Character Set provide shapes useful in constructing straightforward pictures. Such pictures can be displayed in a range of 16 colours. Graphics mode must be used to construct graphic images requiring resolution of display points smaller than provided by text character positions; however, there are fewer colour choices for such images.

3.1.3 Medium-Resolution Graphics Mode

Using the Color/Graphics Monitor, medium-resolution graphics mode can be selected. In this mode, text characters can be displayed in a fashion similar to text mode with the width set to 40. As well, graphic images can be drawn and coloured using functions and subroutines described in this section.

Arcs, lines and other graphic images are drawn with reference to points on the screen, and can be set to one of four colors. Since one of these four colour choices corresponds to the background colour, only three visible colour choices are available. There are two sets of three visible colours with which graphic images may be drawn: green, red and brown; or, cyan, magenta and white. These sets are known as the *palettes* and either one set or the other can be selected. If the palette is changed, all graphic images currently on the screen will change colours to corresponding colours in the new palette.

The background can be set to one of 16 unique colours in medium mode. When the background colour is changed in medium mode, the entire background changes. The foreground colour may be set to one of the four colours of the current palette.

Colours which may be selected in medium mode are described in the section entitled "Colour Choices".

In medium mode, there are 40 x 25 (that is, 1000) character positions for text display. This same area is viewed as 320 x 200 (that is, 64000) points which can be individually set to coloured dots. Series of coloured dots collectively form graphic images. The positions of points (dots) are specified by coordinates as described in the section entitled "Display Coordinates".

3.1.4 High-Resolution Graphics Mode

Using the Color/Graphics Monitor, high-resolution graphics mode can be selected. In this mode, text characters can be displayed in a fashion similar to text mode with the width set to 80. As well, graphic images can be drawn in white on a black background. The screen area has twice the number of points available as in medium mode giving the capability to define crisper images with smoother arcs.

In high mode, there are 80 x 25 (that is, 2000) characters positions for text display. This same area is viewed as 640 x 200 (that is, 128000) points which can be individually set to white or invisible (black) dots. Series of white dots collectively form graphic images. The positions of points (dots) are specified by coordinates as described in the section entitled "Display Coordinates".

3.1.5 Monochrome Display Attributes

Values 0 through 31 can be specified as the foreground "colour" using the Monochrome Display. Background colour can be specified either as 0 (black) or 7 (white). With a background of 0 (black), the 32 foreground colours result in the character attributes shown in the following table.

Colour	Character Attribute
0	black (invisible)
1	underlined white
2-7	white
8	black (invisible)
9	underlined bright white
10-15	bright white
(+16)	blinking colours
16-31	blinking with same "colours" as 0-15

With a background of 7 (white), 4 foreground colours result in the character attributes shown following:

Colour	Character Attribute
0	black (reverse image)
8	light black (reverse image)
16	blinking black (reverse image)
24	blinking light black (reverse image)

All other foreground values result in a *black* background, regardless of the background setting, and yield character attributes as shown in the first table.

3.1.6 Colour Choices

This section describes possible values for *COLOUR*, *PALETTE* and *PCOLOUR*, as referenced in the functions and subroutines described subsequently in this section. Appropriate values for these parameters in text, medium and high modes are illustrated in the following tables. Information in this section applies to the use of the Color/Graphics Monitor.

Text Mode and Medium Background

colour	basic colours	text background	text border	text foreground	medium background
0	black	•	•	•	•
1	blue	•	•	•	•
2	green	•	•	•	•
3	cyan	•	•	•	•
4	red	•	•	•	•
5	magenta	•	•	•	•
6	brown	•	•	•	•
7	white	•	•	•	•
(+8)	light colours				
8	gray		•	•	•
9	light blue		•	•	•
10	light green		•	•	•
11	light cyan		•	•	•
12	light red		•	•	•
13	light magenta		•	•	•
14	yellow		•	•	•
15	bright white		•	•	•
(+16)	blinking colours				
16 through 31	blinking same colours as 0-15			•	

Medium Resolution

palette	pcolour	meaning	graphic image	foreground
0 (even value)	0	background	•	•
	1	green	•	•
	2	red	•	•
	3	brown	•	•
1 (odd value)	0	background	•	•
	1	cyan	•	•
	2	magenta	•	•
	3	white	•	•

High Resolution

pcolour	meaning	graphic image	background	foreground
0	black	•	•	
1	white	•		•

3.1.7 Display Coordinates

For the purpose of text-character display, the screen is treated as 25 lines of 80 or 40 characters (depending upon mode and width settings). The 2000 or 1000 character positions can be referenced by coordinate position (x,y) as illustrated in the following diagram.

```
          Character Coordinates


      (screen top)

                        width 80    width 40

0,0           x           79,0   or   39,0
    .               .



y



    .               .
0,24                      79,24  or   39,24
```

For the purpose of graphics display, a Color/Graphics Monitor is treated as 64000 or 128000 *points* (*dots* when coloured) in medium- and high-resolution modes, respectively. These points can be referenced by coordinates (x,y) as illustrated in the following diagram.

```
         Graphic Point Coordinates


      (screen top)

                        medium      high

0,0           x          319,0   or  639,0
    .               .



y



    .               .
0,199                    319,199 or  639,199
```


Each graphics subroutine or function, upon completion, leaves the "paint brush" at a defined *current point*. Subsequent images drawn can be positioned relative to the current point using the *IXPNT* or *IYPNT* functions described in this section.

One might intuitively expect a box to appear square if drawn with an equal height and width in terms of points. However, the shape will appear rectangular, particularly in high-resolution mode. This is the result of a difference in physical spacing of points between the vertical and horizontal axes. The difference is accentuated in high-resolution mode, since the same horizontal physical width represents twice as many points whereas the number of points in the vertical axis remains constant at 200.

Scaling of coordinates can correct for these differences to present appropriate visual shapes. In both medium and high-resolution modes, points in the vertical (y) direction are further apart than in the horizontal (x) direction. Scaling the number of vertical points by the fractions 5/6 (medium) and 5/12 (high) will approximate a physical length equal to that of the unscaled number of points in the horizontal direction. These fractions are derived as follows:

(1) the physical dimensions of the screen display area are proportionately 4 units by 3 units (for example, 8 inches wide by 6 inches high);

(2) the physical width (4) represents 320 or 640 points, implying 80 or 160 points per unit horizontal spacing;

(3) the physical height (3) represents 200 points, implying 67 points per unit vertical spacing;

(4) 67/80 approximates 5/6 and 67/160 approximates 5/12.

The *CIRCLE* subroutine automatically scales the radius in the vertical (y) direction. The *ARC* subroutine, however, permits direct user control over scaling of elliptical curves by specification of axis length in terms of points. An arc drawn in medium mode with an y-axis which is 5/6 the length of the x-axis will produce a visual circle. Other graphic drawing subroutines and functions require the user to scale coordinates and lengths appropriately as desired.

3.1.8 Display Support Subroutines and Functions

A set of intrinsic subroutines and functions are provided to support colour, graphics and other display attributes for the IBM Personal Computer.

The following descriptions assume that the reader is familiar with the concepts and terminology described thus far. The terms *COLOUR*, *PCOLOUR* and *PALETTE* used in these descriptions are detailed in the sections entitled "Monochrome Display Attributes" and "Colour Choices". Point (dot) coordinates (referenced as *X*, *Y*, *X1*, *Y1*, *X2*, *Y2*) are described in the section entitled "Display Coordinates".

Some of the parameters to the functions and subroutines are optional. These optional parameters are shown in the formal parameter lists enclosed in square brackets: [optional-parameter]. Also, parameters are positional in the respective lists. Where two or more optional parameters may be specified, FORTRAN requires that the optional parameters preceeding the specified parameter must also be specified. In other words, no parameter may be omitted to the left of the parameter in question. In the routines where an optional *PCOLOUR* may be specified, the INTEGER value −1 may be used (since this value does not correspond to any legal colour value). For example,

```
CALL CIRCLE( X, Y, RADIUS, -1, START, END )
```

will draw a circular arc using the default foreground colour. Commas must not appear for unspecified optional parameters at the end of a list. This is illustrated by the following invocation which produces a complete circle with colour value 1.

```
CALL CIRCLE( X, Y, RADIUS, 1 )
```

Unless otherwise stated, the type of the parameters to the graphics subroutines and functions is INTEGER.

Subroutine: ARC(X, Y, X-AXIS, Y-AXIS, [PCOLOUR], [START], [END])

This subroutine is used to draw an ellipse or elliptical arc with centre at coordinates *X*, *Y* and colour corresponding to the *PCOLOUR* value. The *X-AXIS* and *Y-AXIS* determine the shape of the ellipse and their length is expressed in terms of the number of points.

Partial arcs can be drawn by specifying *START* and *END* angles expressed in degrees between 0 and 360. Upon completion, the current point corresponds to the end point on the arc. The default *PCOLOUR* value is the foreground colour. The default values for *START* and *END* are 0 and 360, respectively. If start and end values are specified then the pcolour must be specified.

Subroutine: BACKGR(COLOUR)

This subroutine is used to set the background to the *COLOUR* corresponding to the numeric parameter value. The current point position is unaffected.

Subroutine: BOXFUL(X1, Y1, X2, Y2, [PCOLOUR])

This subroutine is used to draw a rectangular box with coordinates of opposite corners specified by the first two pairs of parameters. The box and area inside of it are painted solidly the colour corresponding to the *PCOLOUR* value. The default *PCOLOUR* value is the foreground colour. Upon completion, the current point has an x-coordinate equal to the minimum of *X1* and *X2*, and has a y-coordinate equal to the minimum of *Y1* and *Y2*.

Subroutine: BORDER(COLOUR)

This subroutine is used to set the border to the *COLOUR* corresponding to the numeric parameter value. The current point position is unaffected.

Subroutine: BOX(X1, Y1, X2, Y2, [PCOLOUR])

This subroutine is used to draw a rectangular box with coordinates of opposite corners specified by the first two pairs of parameters. The box colour corresponds to the *PCOLOUR* value. The default *PCOLOUR* value is the foreground colour. Upon completion, the current point has an x-coordinate equal to the minimum of *X1* and *X2*, and has a y-coordinate equal to the minimum of *Y1* and *Y2*.

Subroutine: CIRCLE(X, Y, RADIUS, [PCOLOUR], [START], [END])

Draw a circle or circular arc with centre at coordinates *x*, *y* and colour corresponding to the *PCOLOUR* value. The *RADIUS*, specified in terms of the number of points, determines the horizontal radius of the arc. The vertical height is automatically scaled to present the visual image of a circle (see the discussion of scaling in the section entitled "Display Coordinates").

Partial arcs can be drawn by specifying *START* and *END* angles expressed in degrees between 0 and 360. Upon completion, the current point corresponds to the end point on the arc. The default values for *START* and *END* are 0 and 360, respectively. If *START* and *END* are specified then the *PCOLOUR* value must also be specified. The default *PCOLOUR* value is the foreground colour.

Subroutine: CLEAR

This subroutine is used to clear the screen to the background colour. The current point or current cursor position is reset to "home" (0,0).

Subroutine: DRAW(CHARACTER-STRING)

This subroutine is used to draw a graphic image specified by the *draw commands* contained in the CHARACTER variable or expression, *CHARACTER-STRING*.

The draw commands are interpretively executed from left to right. Most draw commands move the current point a number of points, *n*, from the current position. Points between the current point and the destination point are coloured to form a graphic image unless the command is prefixed by the single-letter command B. That is, a B-prefix inhibits drawing. Upon completion of each move command, the current point is the destination point, unless the move command is prefixed by the single-letter command N. An N-prefix causes the current point to remain unchanged. No automatic scaling is performed on the number of points specified in move commands. The following table shows single-letter draw commands which are followed by INTEGER numeric values specified as digits.

```
Command   Meaning

U n       move up n points
D n       move down n points
L n       move left n points
R n       move right n points
E n       move diagonally up n points,
                       right n points
F n       move diagonally down n points,
                         right n points
G n       move diagonally down n points,
                          left n points
H n       move diagonally up n points,
                        left n points
C pcolour set graphic image colour
                  to pcolour value
```


A more general move command

```
M p,q
```

specifies movement to a point with coordinates determined by values p and q. If p or q consists solely of digits, it specifies an absolute coordinate. If either is prefixed by a + or − symbol, it represents a relative adjustment to the corresponding coordinate. For example, the statement

```
DRAW( "R10D5L10U5M+10,+5" )
```

will draw a box, 10 points wide by 5 points high, and draw a diagonal line from the starting corner to the opposite corner.

Subroutine: FOREGR(COLOUR)

This subroutine is used to set the foreground to the *COLOUR* corresponding to the numeric parameter value. The current point position is unaffected.

Subroutine: GETPIC(PIC, X1, Y1, X2, Y2)

This subroutine returns a CHARACTER value in the argument *PIC*. The coordinates *X1,Y1* and *X2,Y2* define the opposite corners of a rectangle. The dimensions (shape) of the rectangle and the colour values corresponding to each point within this rectangle are stored in the character string. In medium resolution mode, colour values are stored 4 points per byte. In high resolution mode, colour values are stored 8 points per byte. The resulting string, *PIC*, may be used in subsequent calls to the subroutine *PUTPIC*. The current point position is unaffected.

Function: IGTDOT(X, Y)

This function returns an INTEGER value representing the colour of the point specified by coordinates *X,Y*. Upon completion, the current point position is set to *X,Y*.

Function: IMODE([MODE])

This function returns an INTEGER value representing the current display mode. This value can be one of:

```
0  text
1  medium resolution
2  high resolution
```

The optional parameter may be specified to change the mode setting. The returned value represents the setting prior to such a change. Changing the mode setting will clear the screen and reset the current point or cursor position to "home" (0,0). The background and foreground colours are set to black and white, respectively.

Function: `IWIDTH( [width] )`

This function returns an INTEGER value representing the current display width. This value can be 40 or 80 and represents the number of columns of text characters displayed. The optional parameter may be specified to change the width setting. The returned value represents the setting prior to such a change. A change to the width will clear the screen and reset the current point or cursor position to "home" (0,0).

The *IWIDTH* function will have no effect in high and medium modes, since these modes have fixed widths of 80 and 40 respectively. Similarly, the *IWIDTH* function will have no effect with the Monochrome Display, since it is always of width 80.

Function: `IXPNT( [X] )`

This function returns an INTEGER value corresponding to the x-coordinate of the current point. The optional parameter may be specified to change the current point to a new x-coordinate position, leaving the y-coordinate unchanged. The returned value represents the x-coordinate prior to such a change.

Function: `IYPNT( [y] )`

This function returns an INTEGER value corresponding to the y-coordinate of the current point. The optional parameter may be specified to change the current point to a new y-coordinate position, leaving the x-coordinate unchanged. The returned value represents the y-coordinate prior to such a change.

Subroutine: `LINE( x1, y1, x2, y2, [pcolour] )`

This subroutine may be used to draw a line from the point with coordinates *X1,Y1* to the point with coordinates *X2,Y2*. The colour of the line will correspond to the specified value of *PCOLOUR*. The default *PCOLOUR* value is the foreground colour. Upon completion, the current point has coordinates *X2,Y2*.

Subroutine: `PAINT( X, Y, [PCOLOUR], [BOUNDARY] )`

This subroutine may be used to fill solidly an area of the screen with the colour corresponding to the specified *PCOLOUR* value. Starting at the point *X,Y* colour surrounding points within an area bounded by the edges of the screen or a figure of colour different than the original colour of point *X,Y*. If the optional parameter, *BOUNDARY*, is specified, the paint area is bounded by the edges of the screen or a figure of colour *BOUNDARY*. The current point position is unaffected.

Painting of complex figures can consume large amounts of system stack space. If *PAINT* terminates with a stack overflow, the language processor can be terminated and restarted with a larger system stack area (as described in the section "Invoking WATCOM FORTRAN").

Subroutine: `PALETT( PALETTE )`

This subroutine may be used to set the graphics palette according to the parameter *PALETTE*. The only meaningful values are 0 or 1. The current point position is unaffected.

Subroutine: `PUTDOT( X, Y, [PCOLOUR] )`

This subroutine may be used to place a dot, with colour corresponding to the specified *PCOLOUR* value, at the point with coordinates *X,Y*. The default *PCOLOUR* value is the foreground colour. Upon completion, the current point position is set to *X,Y*.

Subroutine: `PUTPIC( X, Y, CHARACTER-VARIABLE, [ACTION] )`

This subroutine may be used to display the graphic image previously stored in the CHARACTER variable (parameter *CHARACTER-VARIABLE*). See the description of function *GETPIC* for details of the specification of *CHARACTER-VARIABLE*.

The picture is positioned with coordinates *X,Y* designating the top left corner of the rectangle. Parameter *ACTION* indicates how the new picture blends with what is currently in the rectangular area of the screen. Valid values for *ACTION* are described below. The tables illustrate the results of blending current and stored point pcolour values for the different choices of *ACTION*. The default value for *ACTION* is 0. The current point position is unaffected.

action 0: Replace the current picture.

medium stored points

		0	1	2	3
current	0	0	1	2	3
points	1	0	1	2	3
	2	0	1	2	3
	3	0	1	2	3

high stored points

		0	1
current	0	0	1
points	1	0	1

action 1: Replace the current picture with the negative image of the stored picture.

medium stored points

		0	1	2	3
current	0	3	2	1	0
points	1	3	2	1	0
	2	3	2	1	0
	3	3	2	1	0

high stored points

		0	1
current	0	1	0
points	1	1	0

action 2: In high mode, invert current points if their corresponding stored points are non-zero. That is, the result corresponds to the logical exclusive-OR (XOR) of the current and stored points. An image can be superimposed on an existing picture with one XOR invocation of *PUTPIC*, and the existing picture restored with a second, identical invocation of *PUTPIC*. This is useful for animation.

medium stored points

		0	1	2	3
current	0	0	1	2	3
points	1	1	0	3	2
	2	2	3	0	1
	3	3	2	1	0

high stored points

		0	1
current	0	0	1
points	1	1	0

action 3: In high mode, illuminate points only if corresponding points in both current and stored pictures are non-zero. That is, the result corresponds to the logical AND of the current and stored points.

medium stored points

		0	1	2	3
current	0	0	0	0	0
points	1	0	1	0	1
	2	0	0	2	2
	3	0	1	2	3

high stored points

		0	1
current	0	0	0
points	1	0	1

action 4: Superimpose the stored graphic image onto the current graphic image. That is, the result corresponds to the logical OR of the current and stored points.

medium stored points

		0	1	2	3
current	0	0	1	2	3
points	1	1	1	3	3
	2	2	3	2	3
	3	3	3	3	3

high stored points

		0	1
current	0	0	1
points	1	1	1

3.2 Other Intrinsic Functions and Procedures

This section describes additional intrinsic subroutines and functions provided to support specific features of the IBM Personal Computer.

Subroutine: DATE(RESULT)

This subroutine returns, in the CHARACTER variable *RESULT*, a character string in the format "YY/MM/DD" where "YY" represents the year number modulo 100, "MM" represents the month number and "DD" represents the day number.

Function: `ISEG( [NEW-SEG] )`

This function returns an INTEGER value representing the current default segment address. The INTEGER parameter *NEW-SEG* is optional and, if specified, resets the default segment address. This segment address is used as the origin for the address parameters to functions *PEEKB*, *PEEKI*, *POKEB*, *POKEI* and *SYS*.

The current default segment address is reset to the Intel 8088 microprocessor Data-space Segment (DS) value for the microFORTRAN interpreter's workspace each time a program is run. This workspace contains the user's program and data. The value returned by the intrinsic function *IVARPT*, is the address of a user-defined variable in that data-space segment.

Segment addresses for the Intel 8088 microprocessor used in the IBM Personal Computer are, by definition, 16-bit unsigned integers. A segment address defines the origin of a logical address space which is accessed with a 16-bit logical address between 0 and 65535. The real physical memory address of the beginning of a segment is computed by multiplying the segment address by 16 (bytes).

Function: `ISYS( ADDR, [P1] [,P2] . . . [P8] )`

This function may have from one to nine arguments. The mandatory argument, *ADDR*, must be an INTEGER value which is the address within the current *segment* of a routine to be invoked. The current segment is set by the ISEG intrinsic function. The other, optional arguments are passed to the invoked routine as arguments in the following manner:

(1) INTEGER arguments are passed as two-byte integers.

(2) REAL arguments are copied to a temporary memory location and the address of the memory location is passed.

(3) CHARACTER arguments are copied to a temporary memory location, an extra character with hexadecimal value 0 is appended to the string, and the address of the memory location is passed.

(4) All other arguments are illegal. The arguments to be passed to the invoked routine are pushed on the the stack, in order, starting with the last and proceeding to the second. The first argument to be passed is loaded into the AX register of the Intel 8088.

When the invoked routine returns to the WATCOM FORTRAN system, the contents of the AX register is returned to the program as an INTEGER value.

Subroutine: `PAGE( ACTIVE-PAGE, VISUAL-PAGE )`

The subroutine *PAGE* permits the user to set both the *ACTIVE* page and the *VISUAL* page. *ACTIVE-PAGE* and *VISUAL-PAGE* are INTEGER arguments. The active page is the page that is written to by output statements to the screen. The visual page is the page that is currently displayed on the screen. This feature is valid in text mode only when using the Colour/Graphics Monitor. There are 8 possible values in the range 0 to 7 for a screen width of 40 characters per line. There are 4 possible values in the range 0 to 3 for a screen width of 80 characters per line.

The position of the text cursor remains unchanged when changing from one active page to the next. Therefore care should be taken to position the cursor in an appropriate place using output statements.

Care should also be taken to make both the active page and the visual page identical before terminating execution or when soliciting input from the keyboard.

Subroutine: `PFKEY( OLD-KEY, KEY-NUMBER, [KEY-STRING] )`

This subroutine returns a CHARACTER value, in the CHARACTER variable OLD-KEY, that corresponds to the current setting for a user defined program function key. *KEY-NUMBER* is an INTEGER value in the range 1 through 40. Up to 40 unique program function keys may be set. *KEY-STRING* is a CHARACTER expression that represents the character sequence that the user wishes to associate with a particular program function key. The character sequence may contain any characters from the IBM Personal Computer screen character set. This includes the carriage return character (CHAR(13)) among others.

Once a character sequence is assigned to a particular program function key, the depression of that key will cause the character sequence to be echoed on the screen (just as if it had been typed on the keyboard).

This subroutine always returns the previous setting of the corresponding program function key. A null character string is returned if no previous assignment has been made. Settings will remain in effect as long as microFORTRAN is being used. User defined program function keys are active during the execution of the user's program.

Subroutine: `TIME( RESULT )`

This subroutine returns, in the CHARACTER variable *RESULT*, a character string in the format "HH:MM:SS.TT" to represent the hours (HH), minutes (MM), seconds (SS) and sixtieths of a second (TT) of the current time.

Function: **SLEEP(X)**

This function causes the program to pause for a period of time determined by the REAL argument *X*. This time is represented in 60-ths of seconds. When the program resumes execution, the current time of day is returned as the number of 60-ths of seconds since midnight.

Function: **TOD( )**

This function returns the current time of day as a REAL number. The time is represented as the number of 60-ths of seconds since midnight.

APPENDIX B

WATCOM Reference Card
GRAPHICS

WATCOM Language Features Specific to IBM PC

- Colour graphics support.
- 8087 numeric processor used if present.
- Can access files on other systems with Waterloo PC Network and Waterloo HOSTCM communications support.

Additional BASIC Functions

j%=segment[(i%)] — j% is set to the current segment address. If i% is specified, the current segment is changed to i%. This value is used whenever machine addresses are specified (i.e. the PEEK and USR functions and the POKE and SYS statements).

t$=setup$(s$) — Set up the port specified by the first entry of s$ to the parameters specified by the rest of s$. The return string is either an error message, or the status of the specified port if no error occurred.

j%=in(port%) — Data read from the specified port is returned.

call out(port%,i%) — The byte i% is output to specified port.

As noted above (under SEGMENT), the functions PEEK and USR and the statements POKE and SYS use the current segment value when calculating addresses.

Graphics functions and procedures are outlined elsewhere on this reference card.

Additional BASIC Reserved Words

arc	foreground	palette
background	getdot	putdot
block	getpic	putpic
border	in	segment
box	line	setup$
circle	mode	width
clearscreen	out	xpoint
draw	paint	ypoint

FORTRAN Functions and Subroutines

var = iseg([integer expression])
: The current segment address is assigned to var. If a parameter is specified, the current segment is changed to the passed value. This value is used whenever machine addresses are specified (ie the ISYS, IPEEKB, IPEEKI, IPOKEB, and the IPOKEI functions).

var = isys(addr [,p1] [,p2] ... [,p8])
: Invoke the machine code at the address ADDR in the current code segment. One to nine arguments may be passed. The return value is assigned to var.

call pfkey(old-key, key-number [,key-string])
: Define old-key with the value of the PF key referenced by key-number, optionally setting the PF key with the value of key-string.

Pascal Predefined Functions

segment[(i)] : integer — **Segment** returns an integer value representing the current segment address. If a parameter is specified, the current segment is changed to that value. This value is used as the origin for the address parameters to procedures **poke** and **sysproc** and functions **peek** and **sysfunc**.

Graphics functions and procedures are outlined elsewhere on this reference card.

Graphics Functions and Procedures

ARCS Draw an ellipse or elliptical arc with centre at coordinates (x,y) and colour corresponding to the pcolour value.

BASIC
```
call arc( x, y, xaxis, yaxis
    [,pcolour [,start [,end] ] ] )
```
FORTRAN
```
call arc( x, y, xaxis, yaxis
    [,pcolour [,start [,end] ] ] )
```
Pascal
```
arc( x, y, xaxis, yaxis
    [,pcolour [,start [,end] ] ] )
```

CIRCLES Draw a circle or circular arc with centre at coordinates x,y and colour corresponding to the pcolour value.

BASIC
```
call circle( x, y, radius
    [,pcolour [,start [,end] ] ] )
```
FORTRAN
```
call circle( x, y, radius
    [,pcolour [,start [,end] ] ] )
```
Pascal
```
circle( x, y, radius
    [,pcolour [,start [,end] ] ] )
```

COLOUR, BACKGROUND Set the background to the colour corresponding to the numeric parameter value.

BASIC	`call background( colour )`
FORTRAN	`call backgr( colour )`
Pascal	`background( colour )`

COLOUR, BORDER Set the border to the colour corresponding to the numeric parameter value.

BASIC	`call border( colour )`
FORTRAN	`call border( colour )`
Pascal	`border( colour )`

COLOUR, FOREGROUND Set the foreground to the colour indicated.

BASIC	`call foreground( colour )`
FORTRAN	`call foregr( colour )`
Pascal	`foreground( colour )`

DISPLAY, CLEARING Clear the screen to the background colour.

BASIC	`call clearscreen`
FORTRAN	`call clear`
Pascal	`clearscreen`

DISPLAY, MODE Get/replace the current display mode. This value can be one of

```
0 - text;
1 - medium resolution;
2 - high resolution.
```

BASIC	`j%=mode[(i%)]`
FORTRAN	`j=imode([i])`
Pascal	`j:=mode[(i)]`

DISPLAY, WIDTH Get/replace the current display width.

BASIC	`j%=width[(i%)]`
FORTRAN	`j=iwidth([i])`
Pascal	`j:=width[(i)]`

DRAW Draw a graphic image specified by draw commands contained in the string-valued expression.

BASIC	`call draw( string$ )`
FORTRAN	`call draw( string )`
Pascal	`draw( string )`

Command	Meaning
U n	move up n points
D n	move down n points
L n	move left n points
R n	move right n points
E n	move diagonally up n points, right n points
F n	move diagonally down n points, right n points
G n	move diagonally down n points, left n points
H n	move diagonally up n points, left n points
C pcolour	set graphic image colour to pcolour value
M x,y	move to position (x,y)
M sx,sy	s can be '+' or '-' move (sx,sy) units from current cursor position

LINES Draw a line from (x1,y1) to the point with coordinates (x2,y2).

BASIC	`call line( x1, y2, x2, y2 [,pcolour] )`
FORTRAN	`call line( x1, y1, x2, y2 [,pcolour] )`
Pascal	`line( x1, y1, x2, y2 [,pcolour] )`

PAINTING Fill an area of the screen with the colour corresponding to the specified pcolour value, starting at the point (x,y).

BASIC	`call paint( x, y [,pcolour [,boundary] ] )`
FORTRAN	`call paint( x, y [,pcolour [,boundary] ] )`
Pascal	`paint( x, y [,pcolour [,boundary] ] )`

PALETTE SELECTION Set the palette to 0 or 1 as specified by the parameter.

BASIC	`call palette( pal )`
FORTRAN	`call palett( pal )`
Pascal	`palette( pal )`

WATCOM PC Graphics

PICTURES, READING Capture the picture bounded by (x1,y1) and (x2,y2) into the string variable.

BASIC	`call getpic( x1, y1, x2, y2, variable$ )`
FORTRAN	`call getpic( variable, x1, y1, x2, y2 )`
Pascal	`getpic( x1, y1, x2, y2, variable )`

PICTURES, WRITING Display the graphic image previously stored by procedure GETPIC in the string variable at (x,y).

BASIC	`call putpic( x, y, variable$ [,action] )`
FORTRAN	`call putpic( x, y, variable [,action] )`
Pascal	`putpic( x, y, variable [,action] )`

Actions	
0	Set
1	Reset
2	Xor
3	And
4	Or

	Stored Point					
SET	medium 0	1	2	3	high 0	1
0	0	1	2	3	0	1
Scrn 1	0	1	2	3	0	1
Colr 2	0	1	2	3		
3	0	1	2	3		

	Stored Point					
RESET	medium 0	1	2	3	high 0	1
0	3	2	1	0	1	0
Screen 1	3	2	1	0	1	0
Colour 2	3	2	1	0		
3	3	2	1	0		

	Stored Point					
XOR	medium 0	1	2	3	high 0	1
0	0	1	2	3	0	1
Scrn 1	1	0	3	2	1	0
Colr 2	2	3	0	1		
3	3	2	1	0		

	Stored Point					
AND	medium 0	1	2	3	high 0	1
0	0	0	0	0	0	0
Screen 1	0	1	0	1	0	1
Colour 2	0	0	2	2		
3	0	1	2	3		

	Stored Point					
OR	medium 0	1	2	3	high 0	1
0	0	1	2	3	0	1
Scrn 1	1	2	3	3	1	1
Colr 2	2	3	3	3		
3	3	3	3	3		

POINTS, READING Get the colour of the point (x,y).

BASIC	`i%=getdot(x%,y%)`
FORTRAN	`i=igtdot(x,y)`
Pascal	`i:=getdot(x,y)`

POINTS, WRITING Set colour of the point (x,y).

```
BASIC    call putdot( x, y [,pcolour] )
FORTRAN  call putdot( x, y, [pcolour] )
Pascal   putdot( x, y [,pcolour] )
```

POINTS, X-COORDINATE Get/replace the x-coordinate of the current point.

```
BASIC    j%=xpoint[(i%)]
FORTRAN  j=ixpnt([i])
Pascal   j:=xpoint[(i)]
```

POINTS, Y-COORDINATE Get/replace the y-coordinate of the current point.

```
BASIC    j%=ypoint[(i%)]
FORTRAN  j=iypnt([i])
Pascal   j:=ypoint[(i)]
```

RECTANGLES, OUTLINE Draw a rectangular box with coordinates of opposite corners specified by the first two pairs of parameters.

```
BASIC    call box( x1, y1, x2, y2 [,pcolour] )
FORTRAN  call box( x1, y1, x2, y2 [,pcolour] )
Pascal   box( x1, y1, x2, y2 [,pcolour] )
```

RECTANGLES, SOLID Draw a rectangular box with coordinates of opposite corners specified by the first two pairs of parameters. The box and area inside of it are painted the colour corresponding to the pcolour value.

```
BASIC    call block( x1, y1, x2, y2 [,pcolour] )
FORTRAN  call boxful( x1, y1, x2, y2 [,pcolour] )
Pascal   block( x1, y1, x2, y2 [,pcolour] )
```

Monochrome Display Character Attributes

Colour #	Character Attribute
0	black (invisible)
1	underlined white
2 - 7	white
8	black (invisible)
9	underlined bright white
10 - 15	bright white
(+16)	blinking
16 - 31	blinking with same colours as 0-15

With a background of 7 (white), 4 foreground colours result in the character attributes shown following:

WATCOM PC Graphics

Colour #	Character Attribute
0	black (reverse image)
8	light black (reverse image)
16	blinking black (reverse image)
24	blinking light black (reverse image)

All other foreground values result in a *black* background.

Colour Choices

Information in this section applies to use of the Color/Graphics Monitor.

Colours Available in Text Mode and for Background in Medium Mode					
Colour #	Basic Colours	Text Back-ground	Text Border	Text Fore-ground	Medium Back-ground
0	black	•	•	•	•
1	blue	•	•	•	•
2	green	•	•	•	•
3	cyan	•	•	•	•
4	red	•	•	•	•
5	magenta	•	•	•	•
6	brown	•	•	•	•
7	white	•	•	•	•
(+8)	light colours				
8	gray		•	•	•
9	light blue		•	•	•
10	light green		•	•	•
11	light cyan		•	•	•
12	light red		•	•	•
13	light magenta		•	•	•
14	yellow		•	•	•
15	bright white		•	•	•
(+16)	blinking colours				
16 to 31	blinking same colours as 0-15			•	

Colours Available in Medium Mode as Foreground and pcolour values		
Colour #	Meaning in Palette 0 (even palette value)	Meaning in Palette 1 (odd palette value)
0	background colour	background colour
1	green	cyan
2	red	magenta
3	brown	white

High Resolution				
Colour #	meaning	use as pcolour	background	foreground
0	black	•	•	
1	white	•		•

Graphics Point Coordinates		
Resolution	X	Y
medium	0-319	0-199
high	0-639	0-199

WATCOM

415 Phillip Street
Waterloo, Ontario, Canada
N2L 3X2
(519) 886-3700
Telex 06-955 458

PRINTED IN CANADA

Reference Card

WATCOM

PC Graphics

APPENDIX C

WATCOM Reference Card
EDITOR

Function Keys

F1-F10		Shift/F1-F10	
F1 Page Up	F2 Page Down	F11 1/2 Pg Up	F12 1/2 Pg Down
F3 Line Up	F4 Line Down	F13 Fore-ground	F14 Back-ground
F5 Line Insert	F6 Line Delete	F15 Line Split	F16 Line Undelete
F7 Select/ Deselect	F8 Cut	F17 Line Join	F18 Paste
F9 Screen/ Command	F10 PF Key Help	F19 Fill	F20 Edit

Page Up	Moves 20 lines towards the beginning of the workspace.
Page Down	Moves 20 lines towards the end of the workspace.
Line Up	Causes the line before the current line to become the current line.
Line Down	Causes the line following the current line to become the current line.
Line Insert	Adds a blank line after the current line.
Line Delete	Deletes the current line.
Select/Deselect	Begin or cancel selection of a range of lines.
Cut	Move selected lines to the "scrap" workspace.
Screen/Command	Switches the editor between "command" and "screen editing" modes.
PF Key Help	Displays a layout of the function keypad and the functions assigned to each key (includes user-assignable Alt function keys). Press the *Enter* key to resume editing.
1/2 Pg Up	Moves 10 lines towards the beginning of the workspace.
1/2 Pg Down	Moves 10 lines towards the end of the workspace.
Foreground	Steps through a selection of 16 foreground colours.
Background	Steps through a selection of 8 background colours.
Line Split	Splits the current line at the cursor position into two lines.
Line Undelete	Undeletes the line deleted by *Line Delete*.
Line Join	Concatenate the line following the current line to the current line.

WATCOM Editor for the IBM PC

Paste	Copy lines from the "scrap" workspace into the current workspace after the current line.
Fill	Fill the selected lines of text using the margins established by the "fill" command.
Edit	Move to the next workspace.

Numeric Keypad Keys

(Backspace) <—	Num Lock		Scroll Lock (Break)
Enter <┘	7 Home	8 ↑	9 PgUp
	4 ←	5	6 →
PrtSc •	1 End	2 ↓	3 PgDn
0 Ins		Del	

Backspace	erases the character to the left of the cursor. Characters to the right are shifted left. This key is most useful for deleting the last character typed.
Num Lock	is used to switch between the number keys on the numeric keypad and the cursor control and other keys.

Ctrl/Num Lock (*Pause*)
temporarily halts the editor so that you may view the contents of the screen. Press any key (except "Ctrl/Num Lock") to continue.

Ctrl/Scroll Lock (*Break*)
will generally terminate the current editor command. When modifying a line in "screen" mode, this key combination may be used to restore the line back to its unmodified state.

Enter	is used to enter commands to the editor when in command mode. It acts as a carriage return, when in screen mode, moving the cursor to the start of the next line.
Home	causes the screen to be redisplayed with the current line at the "home" position (the top line of the screen).
Ctrl/Home	causes the screen to be redisplayed.
PgUp	moves 20 lines towards the beginning of the workspace.
PgDn	moves 20 lines towards the end of the workspace.
Cursor Up ↑	selects the line previous to the current line as the new current line.

Cursor Down ↓ — selects the line following the current line as the new current line.

Cursor Left ← — moves the cursor one position left.

Cursor Right → — moves the cursor one position right.

Shift/PrtSc — (*Print Screen*) causes the contents of the screen to be printed on the printer.

End — moves the cursor to the last character on the line.

Ctrl/End — erases all characters on the line above and to the right of the cursor.

Ins — will cause the editor to enter character insertion mode. Characters will be inserted at the cursor position. All characters to the right of the cursor will be shifted to the right. Press *Ins* a second time to cancel insert mode.

Del — deletes the character where the cursor is positioned. Characters to the right are shifted left.

Other Keys

Esc — has no function assigned to it.

Tab Forward — used to move the cursor right to the next tab stop.

Tab Backward — used to move the cursor left to the previous tab stop.

Control Keys

Ctrl/B — when used in conjunction with a language interpreter, causes execution of the program to begin.

Ctrl/C — will generally terminate the current editor command. When modifying a line in "screen" mode, this key combination may be used to restore the line back to its unmodified state.

Ctrl/G — produces a sound from the PC.

Ctrl/H — functions identically to the "CURSOR LEFT" key.

Ctrl/I — functions identically to the "TAB" key.

Ctrl/J — functions identically to the "CURSOR DOWN" key.

Ctrl/K — functions identically to the "HOME" key.

Ctrl/L — causes the screen to be rewritten.

Ctrl/M — functions identically to the "ENTER" key.

Ctrl/[— is ignored.

Ctrl/\ — is equivalent to the "CURSOR RIGHT" function.

Ctrl/] — is equivalent to the "CURSOR LEFT" function.

Ctrl/^ — is equivalent to the "CURSOR UP" function.

Ctrl/_ — is equivalent to the "CURSOR DOWN" function.

All other control keys result in special characters appearing on the screen.

Meta Characters

Certain sequences are used to form characters which have special meanings. The following may be used in text searching, matching and substitution.

%/ — represents a "/"

%^ — represents the beginning of a line

%$ — represents the end of a line

%. — represents any character

%* — represents zero or more repetitions of the character which immediately precedes the "%*"

%% — represents a "%"

There are two additional meta-characters which apply only to the replacement text in a **change** command.

%& — represents any text that was matched by the search string of the change command.

%# — represents the line number of the line that is being operated upon.

Line Referencing

A **line-specifier** is one of the following:

 — the current line

n — the *n*th line in the file (0, 1, 2, ...)

$ — the last line in the file

\+ — the line after the current line

+*n* — the *n*th line after the current line

− — the line previous to the current line

−*n* — the *n*th line previous to the current line

/text/ — the first line in the file that contains the string "text"

+/text/ — the first line following the current line that contains the string "text"

−/text/ — the first line preceding the current line that contains the string "text"

A *line-specifier* may also be null, in which case it indicates the current line.

A **line-id** is one of the following:

line-specifier — the line indicated by *line-specifier*

line-specifier+n — the line *n* lines after the line indicated by *line-specifier*

line-specifier−n — the line *n* lines before the line indicated by *line-specifier*

A **line-range** is one of the following:

line-id — the line indicated by *line-id*

line-id,line-id — all lines from the first *line-id* to the second *line-id*. If either *line-id* is omitted then "." (the current line) is assumed.

* — all the lines in the file (equivalent to 1,$).

*/text/ — all lines containing "text".

*line-id^* — all the lines from *line-id–20* up to and including *line-id* (a maximum of 21 lines); the '^' is obtained by pressing SHIFT and the 6 key located in the top row of the keyboard (equivalent to *line-id–20,line-id*).

line-id& — all the lines from *line-id* up to and including *line-id+20* (a maximum of 21 lines); the '&' is obtained by pressing SHIFT and the 7 key located in the top row of the keyboard (equivalent to *line-id,line-id+20*).

Editor Commands

In the following, any items surrounded by '⟨' and '⟩' brackets indicate items that are specified by the user. For example, "⟨line-range⟩" may be substituted with any valid *line-range* specification such as ".,.+10" or simply ",+10". Note that the brackets are not entered.

— displays the current line-number.

= — repeats the last command.

? — recalls the last command so that it may be altered and resubmitted.

@ ⟨file-name⟩ — executes the editor commands stored in the file specified by ⟨file-name⟩

⟨line-range⟩APPend ⟨file-name⟩
copies all the lines specified by ⟨line-range⟩ to the end of the file specified by ⟨file-name⟩.

AUTosave ⟨count⟩
used to enable or disable automatic saving of the current workspace to the file "autosave.tmp" after every ⟨count⟩ changes to the workspace. When ⟨count⟩ is 0, this feature is disabled.

BYE — used to leave the current workspace after any changes have been saved to a file.

CANcel — used to quit from the editor even though changes may not have been saved.

⟨line-range⟩Change/text1/text2/
changes the first occurrence of "text1" to "text2" in the specified line range.

⟨line-range⟩Change*n*/text1/text2/
changes the '*n*'th occurrence of "text1" to "text2" in the specified line range.

⟨line-range⟩Change*/text1/text2/
changes all occurrences of "text1" to "text2" in the specified line range.

⟨line-id⟩Change — positions to the line specified by ⟨line-id⟩ and places the cursor on that line to allow on-screen modification.

⟨line-range⟩CUT ⟨ws-name⟩
moves the lines specified by ⟨line-range⟩ to the named workspace. If ⟨ws-name⟩ is not specified then ".scrap." is used.

DATe — displays the current date and time.

⟨line-range⟩Delete
deletes all lines in the specified line range.

DIrectory ⟨directory-name⟩ ⟨file-name⟩
lists the names of all the files that reside in the specified directory. If an optional file name is specified then the directory listing is placed in that file.

EDit ⟨file-name⟩ — used to edit a new or old workspace. For a new workspace, the file specified by ⟨file-name⟩ is copied into the workspace provided that the file exists. If ⟨file-name⟩ is not specified, the editor moves to the next workspace, if any; otherwise it remains in the current workspace.

ERAse ⟨file-name⟩
causes the named file(s) to be deleted.

EXit ⟨file-name⟩ — copies all the lines in the workspace into the file specified by ⟨file-name⟩. If a filename is not specified then the current workspace name is assumed. If the file is written successfully, the editor moves to the next workspace, if any; otherwise it terminates.

⟨line-range⟩Fill ⟨lm⟩ ⟨rm⟩ ⟨indent⟩
(Editor only) causes the text within the specified line range to be "filled out" between the left and right margins. A positive or negative indentation for the first line in the resulting "filled" lines may be specified.

⟨line-id⟩Get ⟨file-name⟩
causes the entire contents of the specified file to be inserted following the line specified by ⟨line-id⟩.

Help ⟨topic⟩ — causes help for the topic ⟨topic⟩ to be displayed. If ⟨topic⟩ is not specified then a list of Editor commands is displayed.

⟨line-id⟩Input — sets the current line to ⟨line-id⟩, and then puts the editor into input mode. Subsequent text is entered into the workspace, until a line containing only a period (".") is entered.

⟨line-range⟩ — Text may be viewed ("listed") simply by specifying the range of text to be listed.

MARgin ⟨mar⟩ — sets the left margin for viewing lines in the workspace.

Msg ⟨message⟩ — replaces the bottom line of the screen with the text of ⟨message⟩.

Name ⟨ws-name⟩ — sets or displays the current workspace name which the **put** and **exit** commands use as the default filename. When ⟨ws-name⟩ is not specified the current workspace name is displayed.

PAGe ⟨ps⟩ ⟨tm⟩ ⟨bm⟩ ⟨in⟩
(Editor only) used in conjunction with the "PRInt" command. ⟨ps⟩ sets page size; ⟨tm⟩ sets top margin; ⟨bm⟩ sets bottom margin; ⟨in⟩ sets left margin indentation.

⟨line-id⟩PASte ⟨ws-name⟩
: causes the workspace specified by ⟨ws-name⟩ to be copied after the line specified by ⟨line-id⟩. If ⟨ws-name⟩ is not specified then ".scrap." is used.

PF ⟨number⟩ ⟨text⟩
: used to assign a string of ⟨text⟩ to Alt function keys 1 through 10. The function key number is specified by ⟨number⟩.

⟨line-range⟩PLay ⟨begin⟩
: (Editor only) plays the music in the specified line range. The default ⟨line-range⟩ is the entire workspace. If the keyword "**begin**" is specified then all music defaults are reset.

⟨line-range⟩PRInt ⟨file-name⟩
: (Editor only) prints all the lines specified by ⟨line-range⟩ to the file specified by ⟨file-name⟩ using the parameters established by the "PAGe" command. The default ⟨line-range⟩ is the entire workspace. If ⟨file-name⟩ is not specified then "printer" is assumed.

⟨line-range⟩Put ⟨file-name⟩
: copies all the lines specified by ⟨line-range⟩ into the file specified by ⟨file-name⟩. If the ⟨line-range⟩ is not specified then the entire workspace is copied (equivalent to "*p ⟨file-name⟩"). If a ⟨file-name⟩ is not specified then the current workspace name is assumed.

QUIT
: is used to abandon the current workspace without saving any changes to a file.

REName ⟨oldname⟩ ⟨newname⟩
: causes the file specified by the first filename to have its name changed to the second filename.

RUN
: When the editor is used in conjunction with a language processor, the **run** command executes the program in the current workspace.

SHow
: displays the names of all the workspaces currently being edited.

SYStem ⟨program⟩
: used to invoke IBM PC DOS or some other executable application from within the editor. The editor and workspaces remain in memory. If ⟨program⟩ is specified then an attempt is made to invoke it; otherwise DOS is invoked. The program name must include the file extension.

TIMe
: displays the current time.

WATCOM

415 Phillip Street
Waterloo, Ontario, Canada
N2L 3X2

Phone (519) 886-3700
Telex 06-955 458

Printed in Canada

Reference Card for the IBM PC

WATCOM

Editor

APPENDIX D

WATCOM Reference Card
FORTRAN

V2.0 Features

- Full screen editor is built in.
- Extensive collection of structured programming statements.
- Statements may be entered without regard to columns in a line.
- Long variable names, limited by the length of the line.
- Additional intrinsic routines and a set of intrinsic subroutines for programming flexibility.
- Undefined variable checking (may be turned off).
- Interactive debugger which is automatically invoked when the PAUSE statement is executed or an error is encountered.
- Sequential and direct input/output.
- Subset of the FORTRAN-77 FORMAT codes.

Statement Summary

The following short forms are used throughout the summary:

iconst	integer constant
ivar	integer variable
irvar	integer or real variable
irexpr	integer or real expression
stmnt#	statement number
identifier	character sequence which makes up a name
array_declarator	identifier(iconst [,iconst ...])
var-array	simple variable or array declarator
cexpr	character expression
condition	an integer expression that evaluates to TRUE if non-zero, FALSE if zero. Logical operations such as '<=' result in zero if FALSE, one if TRUE.
cfmt-spec	a fmt-spec which is contained in a character variable, array, or or within a literal string.
fmt-spec	(format-items)

Most format-items are separated by a comma. The 'r' before a format code represents a repetition factor. A 'w' is the field width, a 'd' is the number of decimal places, and a 'p' is a positive integer constant. The following codes are used to cause data to be transmitted.

rA[w]
rIw
rFw.d
rEw.d

The following codes cause data to be inserted.

rX
' ... '
" ... "
rTp
rTLp
rTRp

The following codes affect the transmission of data to and from a record. It is not necessary to use a comma to separate these from other items.

:

example - (i8,2(5x,i4),3(5x,4(i2,2x)),i2)

WATCOM FORTRAN

```
CALL identifier[( [argument [, argument ...] ] ) ]

CHARACTER[clen] var_array[clen]
            [, var_array[clen] ... ]
     where clen is one of:
           •iconst
           •(•)

CLOSE( option [,option ... ] )
     where option is one of:
           unit   = irexpr
           iostat = ivar
           err    = stmnt#

COMMON slist|clist
     where a "slist" is given by
           variable [, variable ... ]
     where a "clist" is given by
           /identifier/slist [/identifier/slist ... ]

CONTINUE

DIMENSION array_declarator [, array_declarator ... ]

DO irvar = irexpr, irexpr [, irexpr ]
    .
    . (body of loop)
    .
ENDDO

DO stmnt# irvar = irexpr, irexpr [, irexpr ]
    .
    . (body of loop)
    .
stmnt# statement

END

EXTERNAL identifier [, identifier ... ]

EXECUTE identifier

FORMAT fmt-spec

[type] FUNCTION identifier
[( [ parameter [, parameter ... ] ] ) ]

   GOTO stmnt#
or GO TO stmnt#

   GOTO(iconst [, iconst ... ] ) [,] expression
or GO TO (iconst [, iconst ... ] ) [,] expression
```

WATCOM FORTRAN

```
GUESS                 [: block id ]
 .
 .
[ADMIT                [: block id ] ]
[ . ]
[ . ]
[ . ]
[ADMIT                [: block id ] ]
[ . ]
[ . ]
[ . ]
ENDGUESS              [: block id ]

IF condition [THEN [DO ] ]           [: block id ]
 .
 .
[ELSEIF condition [THEN [DO ] ]      [: block id] ]
[ . ]
[ . ]
[ . ]
[ELSEIF condition [THEN [DO ] ]      [: block id] ]
[ . ]
[ . ]
[ . ]
[ELSE  [DO ]                         [: block id] ]
[ . ]
[ . ]
[ . ]
ENDIF

IF (expression) [THEN] statement

IF (expression) stmnt#, stmnt#, stmnt#

INTEGER var_array [, var_array ... ]

   LOOP              [: block id ]
or WHILE condition   [: block id ]
    .
    . (body of loop)
    .
   ENDLOOP           [: block id ]
or ENDWHILE          [: block id ]
or UNTIL condition   [: block id ]

OPEN( option [, option ... ] )
     where option is one of:
          unit    = irexpr
          file    = cexpr
          access  = cexpr
          recl    = irexpr
          iostat  = ivar
          err     = stmnt#

PAUSE

PRINT [ [cfmt-spec] [, expression
     [,expression ... ] ] ]
```

```
PROGRAM identifier


QUIT                   [: block id ]


QUITIF condition       [: block id ]


   READ( option [, option ... ] ) [item [, item ... ] ]
or READ [ [cfmt-spec] [, [item [, item ... ] ] ] ]
       where "option" is one of:
              unit    = irexpr
              fmt     = cfmt-spec
              rec     = irexpr
              err     = stmnt#
              end     = stmnt#
              iostat  = ivar


REAL var_array [, var_array ... ]


REMOTE BLOCK identifier
   ... (body of remote block)
ENDBLOCK


RETURN


REWIND( [unit =] irexpr )


STOP


SUBROUTINE identifier
     [( [ parameter [, parameter ... ] ] ) ]


WRITE( option [, option ... ] ) [item [, item ... ] ]
     where "option" is one of:
            unit    = irexpr
            fmt     = cfmt-spec
            rec     = irexpr
            err     = stmnt#
            end     = stmnt#
            iostat  = ivar
```

Operator Priorities

Priority	Operator
8	enclosed in parentheses
7	**
6	*, /
5	+, -
4	//
3	comparisons
2	.AND.
1	.OR.

Options

•option nodefcheck
This option disables the verification that variables have received values before they are referenced in expressions.

•option defchar=number|'character'|"character"
This option defines the character used to verify that values of variables have been defined when they are used in expressions. The default value is 252.

Reserved Words

ACCESS	ENDWHILE	QUIT
ADMIT	ERR	QUITIF
BLOCK	EXECUTE	READ
CALL	EXTERNAL	REAL
CHARACTER	FILE	REC
CLOSE	FMT	RECL
COMMON	FORMAT	REMOTE
CONTINUE	FUNCTION	RETURN
DATA	GO	REWIND
DIMENSION	GOTO	STOP
DO	GUESS	SUBROUTINE
ELSE	IF	THEN
ELSEIF	INTEGER	TO
END	IOSTAT	UNIT
ENDBLOCK	LOOP	UNTIL
ENDDO	OPEN	WHILE
ENDGUESS	PAUSE	WRITE
ENDIF	PRINT	
ENDLOOP	PROGRAM	

Intrinsic Functions

Intrinsic functions are implicitly typed. Functions that start with the letters I through to N return an integer value, while all others return a real value. The only exception is the CHAR function, which returns a character value of length one. The typing of parameters follows the same convention, with character parameters starting with a C.

Function	*Meaning*
ABS(X)	The absolute value of the REAL argument X.
ACOS(X)	The trigonometric ARCCOSINE (in radians) of the REAL argument X, where -1 <= X <= 1 and 0 <= ACOS(X) <= PI.
AINT(X)	The integral portion of the REAL argument X obtained by truncating the fraction part of X.
ALOG(X)	The natural logarithm of the REAL argument X. The argument must be greater than zero.
ALOG10(X)	The logarithm (base 10) of the REAL argument X. The argument must be greater than zero.
AMAX0(I1, I2, ..., IN)	The maximum of the INTEGER arguments I1, I2, ..., IN.
AMAX1(X1, X2, ..., XN)	The maximum of the REAL arguments X1, X2, ..., XN.

AMIN0(I1, I2, ..., IN)
The minimum of the INTEGER arguments I1, I2, ..., IN.

AMIN1(X1, X2, ..., XN)
The minimum of the REAL arguments X1, X2, ..., XN.

AMOD(X1, X2) The REAL modulus of X1, modulo X2, computed as X1 − INT(X1/X2) * X2. X2 cannot be zero.

ANINT(X) The nearest integer value to the REAL argument X.

ASIN(X) The trigonometric ARCSINE (in radians) of the REAL argument X, where −1 <= X <= 1 and −(PI/2) <= ASIN(X) <= (PI/2).

ATAN(X) The ARCTANGENT (in radians) of the REAL argument X, where −(PI/2) <= ATAN(X) <= (PI/2).

ATAN2(X, Y) The ARCTANGENT (in radians) of the quotient X/Y for the REAL arguments X and Y. The value returned is summarized by the following table:

X	*Y*	*Result*
>0	>0	+|arctangent(X/Y)|
>0	=0	PI/2
>0	<0	+|arctangent(X/Y)|
=0	>0	0
=0	=0	*** ILLEGAL ***
=0	<0	PI
<0	>0	-|arctangent(X/Y)|
<0	=0	-PI/2
<0	<0	-|arctangent(X/Y)|

CHAR(I) The I-th character in the collating sequence of the computer. I is an INTEGER argument.

COS(X) The trigonometric COSINE of the REAL argument X, where X is expressed in radians.

COSH(X) The hyperbolic COSINE of the REAL argument X.

DIM(X1, X2) The positive difference between the REAL arguments X1 and X2: if X1 > X2, then (X1 − X2) is returned; otherwise, 0 is returned.

EXP(X) The mathematical constant e raised to the power of the REAL argument X.

FLOAT(I) The value of the INTEGER argument I.

IABS(I) The absolute value of the INTEGER argument I.

ICHAR(C) The position in the collating sequence of the CHARACTER argument.

IDIM(I1, I2) The position difference between the INTEGER arguments I1 and I2: if I1 > I2, then (I1 − I2) is returned; otherwise, 0 is returned.

WATCOM FORTRAN

IFIX(X) The value of the REAL argument X.

INDEX(C1, C2) The position that the CHARACTER argument C2 first occurs in the CHARACTER argument C1. If the contents of C2 are not found anywhere in C1, zero is returned.

INT(X) The integral portion of the REAL argument X.

INTCH(C) The number represented in the CHARACTER argument C.

INTHEX(C) The number represented in the CHARACTER argument C, interpreted as containing hexadecimal digits.

IPEEKB(I) The character in the computer memory located by the value of the INTEGER argument I.

IPEEKI(I) The contents of the INTEGER located in the computer memory located by the value of the INTEGER argument I.

IPOKEB(I1,I2) This function causes the value of the INTEGER argument I2 to be stored in the character in the computer memory located by the value of the INTEGER argument I1. The function returns an INTEGER value representing the previous contents of that memory location.

IPOKEI(I1,I2) This function causes the value of the INTEGER argument I2 to be stored in the INTEGER at the computer memory location located by the INTEGER argument I1. The function returns an INTEGER value representing the previous contents of that memory location.

ISIGN(I1, I2) The absolute value of the first INTEGER argument with the sign of the second INTEGER argument.

ISYS(I, A, A, ..., A)
Invoke a user-written routine that is present outside of the FORTRAN environment. The conventions for receiving the passed parameters is system dependent.

IVARPT(V) The address of the argument V. The argument must be an item which is capable of being assigned a value.

LEN(C) The length of the character string C.

LENSTR(C) The number of the characters that would remain in the string C, if all spaces characters were removed from the end of the string. Zero is returned when the string C consists only of space characters.

LGE(C1, C2) Returns 1 when C1 is greater than or equal to C2; otherwise, 0 is returned.

LGT(C1, C2) Returns 1 when C1 is greater than C2; otherwise, 0 is returned.

WATCOM FORTRAN

LLE(C1, C2) Returns 1 when C1 is less than or equal to C2; otherwise, 0 is returned.

LLT(C1, C2) Returns 1 when C1 is less than C2; otherwise, 0 is returned.

MAX0(I1, I2, ..., IN)
The maximum of the INTEGER arguments I1, I2, ..., IN.

MAX1(X1, X2, ..., XN)
The maximum of the REAL arguments X1, X2, ..., XN.

MIN0(I1, I2, ..., IN)
The minimum of the INTEGER arguments I1, I2, ..., IN.

MIN1(X1, X2, ..., XN)
The minimum of the REAL values X1, X2, ..., XN.

MOD(I1, I2) The modulus of I1, modulo I2, computed as I1−(I1/I2) * I2. I2 cannot be zero.

NINT(X) The nearest integer value to the REAL argument X.

REAL(I) The value of the INTEGER argument I.

REALCH(C) The number represented in the CHARACTER argument C.

RND(X) A random value between 0 and 1. The value is predictably calculated from the value of the REAL argument X, unless this value is 0. When 0 is passed, an unpredictable random number is returned.

SIGN(X1, X2) The value of the first REAL argument with the sign of the second REAL argument.

SIN(X) The trigonometric SINE of the REAL argument X, where X is in radians.

SINH(X) The hyperbolic SINE of the REAL argument X.

SLEEP(X) This function causes the program to pause for a period of time determined by the REAL argument X. When the program resumes execution, the function returns the current time of day in the same manner as the TOD intrinsic function. The values of the return value and of the argument are system dependent.

SQRT(X) The square root of the REAL argument X. The argument must be greater than or equal to zero.

TAN(X) The trigonometric TANGENT of the REAL argument X, where X is expressed in radians.

TANH(X) The hyperbolic TANGENT of the REAL argument X.

TOD() The current time, which is system dependent.

Reference Card

Intrinsic Subroutines

Subroutine	*Meaning*
CHINT(C, I)	A representation of the INTEGER argument I is returned in the CHARACTER argument C.
CHREAL(C, X)	A representation of the REAL argument X is returned in the CHARACTER argument C.
DATE(C)	The current date is returned in the CHARACTER argument C. The format of the character string is system dependent.
HEXINT(C, I)	This subroutine returns, in the CHARACTER argument C, a representation (in hexadecimal format) of the INTEGER argument I.
RPT(C1,C2,I)	This subroutine returns, as a CHARACTER string in argument C1, the string obtained by repeating the CHARACTER argument C2 the number of times indicated by the third INTEGER argument. The third argument must be positive.
SUBSTR(C1, C2, I1, I2)	This subroutine returns, in the CHARACTER argument C1, the string obtained from the CHARACTER string C2, starting at the position indicated by the INTEGER argument I1, for a length indicated by the INTEGER argument I2. This function differs from the substring operation in the following ways: • the argument C2 may be an expression • when argument I1 is non-positive, it is treated as having a value one • when argument I1 is larger than the length of the string C2, it is treated as having that value • when argument I2 is non-positive, it is treated as having a value one • when the number of characters in the string, from the position indicated by I1 onwards, exceeds I2, I2 is treated as being the number of characters remaining.
TIME(C)	This subroutine returns a character string, in the CHARACTER argument C, which represents the current time. The format of the character string is system dependent.

WATCOM

415 Phillip Street
Waterloo, Ontario, Canada
N2L 3X2

Phone (519) 886-3700
Telex 06-955 458

PRINTED IN CANADA

WATCOM

FOR-
TRAN
-OR-
TRAN